ROHM AND HAAS
History of a Chemical Company

Otto Haas
1872–1960

Dr. Otto Röhm
1876–1939

ROHM AND HAAS ‹

HISTORY OF A CHEMICAL COMPANY ‹

Sheldon Hochheiser

upp

University of Pennsylvania Press · Philadelphia · 1986

Designed by Adrianne Onderdonk Dudden

Library of Congress Cataloging in Publication Data

Hochheiser, Sheldon.
  Rohm and Haas: history of a chemical company.

  Includes bibliographies and index.
  1. Rohm and Haas Company—History.   2. Chemical
industry—United States—History.   I. Title.
HD9651.9.R6H63   1985   338.7'66'00973   84-25770
ISBN 0-8122-7940-9

Printed in the United States of America

CONTENTS

LIST OF ILLUSTRATIONS

ACKNOWLEDGMENTS

In the course of researching and writing this book, many people have given me assistance. I am pleased to have the opportunity here to acknowledge their help.

First and foremost thanks must go to John McKeogh, director of corporate communications for the Rohm and Haas Company. It was John's idea that Rohm and Haas should have a proper history written by a professional historian. He gained approval for the project from the appropriate corporate executives, helped me understand the current corporate structure and businesses, and located for me the people who could answer the questions I sometimes didn't even know to ask.

The number of people in every department at Rohm and Haas who helped in some way during the last two years is staggering, and I apologize that I cannot begin to thank them all by name. They were most cooperative in putting their knowledge and resources at my disposal and patient with the peculiarities of an academic loose in the corporate world. Laura Hadden, in particular, deserves thanks for her good sense of the finer points of the English language and for her work in picture selection and caption writing.

Several Rohm and Haas executives supported the project through months when there was no tangible evidence of progress, and then read at least part of the manuscript. Among these were Vincent Gregory, Donald Felley, John Haas, Norman Harberger, and Larry Wilson. I thank them all. Frederick "Fritz" Rarig was a great help in getting me attuned to the company's history. John Llewellyn unearthed boxes of financial records going back to the company's beginning, and helped me make sense out of them. Ruth Reitmeier took a personal interest in the project, and saw to it that my work received prompt attention from her staff of secretaries. Linda Casey most ably handled the often difficult task of transcribing oral history tapes. Sandra Hostetter was invariably patient with requests far from those normally fielded by a business librarian. Eric Meitzner and Margot Licitis translated important German-language documents. My graduate student assistants, first Thom Larson and then Mike Osborne, lifted much of the drudgery of scholarship off my shoulders and served as valuable sounding boards for often half-baked ideas.

Arnold Thackray began advising Rohm and Haas on its history project before I started. He offered innumerable helpful suggestions throughout this project, and helped me retain my perspective as a scholar while I was spending my days in an office at the company. His criticisms of earlier drafts led to many improvements throughout the manuscript. The manuscript also was read by Louis Galambos, George Wise, and Leon Gortler; without their assistance and insightful critiques I would have a far less worthy book. The comments of the two anonymous referees engaged by the Press also proved useful. Tom Rotell, director of the Press, deserves thanks as well.

Finally, I am pleased to have this opportunity to thank Aaron Ihde, my teacher, colleague, and friend. He taught me, by his words and by his example, how to be a historian, and I owe him a debt beyond repayment. He showed me chemistry is not a discipline whose history exists in a vacuum, but a subject at all times interconnected with its surroundings.

April 1985   *Sheldon Hochheiser*
                                        Philadelphia, Pennsylvania

INTRODUCTION

Nineteen eighty-four was the seventy-fifth anniversary of the founding of the Rohm and Haas Company of Philadelphia. For companies, as for people, major anniversaries are milestones to be commemorated. This book began as one such celebration. It is a volume commissioned by Rohm and Haas for publication in conjunction with the anniversary year. It is both my expectation and that of the Rohm and Haas management that this volume will be read widely by the people of Rohm and Haas. If you are one such reader, I hope you will find the narrative which follows interesting, and I will be pleased if it enriches your understanding and appreciation of the enterprise of which you are a part. If this book should refresh memories and answer some long-pondered questions, it will have accomplished an important goal.

But this book is more than just a pleasant souvenir. Many companies put out commemorative publications at significant anniversaries. As often as not, the resulting publications are glossy coffee-table volumes, long on pictures and short on text, leafed through once and then forgotten. Rohm and Haas wanted something more, a real history with lasting value, and this book is the result. It is not just a collection of attractive photographs and amusing anecdotes. This volume is a souvenir of an anniversary, but it is also an attempt at a serious analysis of the history of Rohm and Haas. It strives not merely to retell past events, but to understand them and their context. Historians often claim it is their goal to reach a lay audience at the same time that they are meeting the accepted standards and expectations of their profession. Too often, little more than lip service is paid to this admirable intention. I am convinced it is possible to write a book of value to both a lay audience and the scholarly community. This is an explicit attempt to do so.

This history of the Rohm and Haas Company is important and interesting in and of itself, but no institution exists in a vacuum. Its individual story is a combination of things peculiar to Rohm and Haas and things common to broad strata of our society. Studying its history, therefore, should not only teach us about the company, but also provide lessons on the nature of high-technology industries and their role in the modern American economy.

Over the first half of the twentieth century, the chemical industry has emerged as one of the central components of industry and life. Synthetic chemical products pervade the modern environment: plastics, fibers, coatings, organic pesticides, pharmaceuticals. For the most part, these are new to our century. Many of today's major chemical firms trace their roots to the years before World War I. For instance, Dow started in 1897 and Monsanto in 1901. Du Pont, which for many years has ranked as the largest company in the American industry, goes back a full century earlier to 1802—but it remained a manufacturer of explosives until the 1910s. There is a good reason why this is the case. It was only late in the nineteenth century that chemical knowledge had advanced to the point where it was possible to derive useful new chemical products out of scientific research.

The synthetic dyestuffs industry, which rose mainly in Germany in the last third of the nineteenth century, was the model, the pioneering science-based chemical business. The dye firms, led by such still familiar names as Bayer, Hoechst, and Badische (BASF), worked closely with and supported academic chemists. They hired the universities' graduates to work in some of the first industrial laboratories in the pursuit of new products and new processes. They sold their new dyes to the textile trade and to other customers through technically skilled salesmen who could work in the mills to solve the customers' problems. By 1900 these strategies led the German firms to domination of the world's dyestuffs markets.

Into this tableau stepped the two men of central interest to our story—Dr. Otto Röhm, a chemist graduated from the University of Tübingen, and Mr. Otto Haas, a German native working in the United States. Mr. Haas had experience in the banking and meat exporting businesses, but more importantly, had been employed as a clerk for one of the smaller German dyestuffs firms. What the two men shared, besides friendship, was a desire to become successful on their own. The route they chose, based on their training and experience, was to start their own company based on Dr. Röhm's invention, Oropon bate, for use by leather tanners.

The company that Mr. Haas started when he returned to America in 1909 will be traced in the pages that follow. It is a story of first one man, and then an entire organization, engaged in the management of science and technology for profit. It is a story of an institution that grew alongside both its country and its science, all the while keeping its own unique qualities. It is a story of a man and a company which succeeded for many decades by following a set of central business principles. It is the story of a man who inspired a loyalty that bordered on reverence through his concern for his employees' well-being. (Long after his departure, it remains unthinkable to refer to the founder in terms other than the respectful "Mr. Haas.")

There are several trails, or themes, which will emerge. A desire and an ability to stay at the scientific and technological frontier characterize the company strategy throughout its history, although the sources of new products gradually changed. These sources evolved from a reliance on Dr. Röhm's personal genius, to a broader exploitation of technology transfer in which Mr. Haas took advantage of his roots to obtain products of the wider German chemical industry, to development of a strong, indigenous research division. A parallel evolution can be found in the structure of the company organization, and in the way responsibilities were divided and organized among Rohm and Haas management. For fifty years the company remained a personal, entrepreneurial enterprise. As the company grew in size, however, Mr. Haas had no choice but to delegate more authority. After 1960 the company evolved a more rational structure of the sort befitting the large, multinational corporation. In part, this organization was typical of most companies of corresponding size and milieu, but it was also an individual answer to the particular problems facing the second and third generations of Rohm and Haas management.

Mr. Haas's successors may have changed the structure of the firm, but for the most part they have found that success continued to flow from a central core of business strategies which they have inherited. These principles, including continued development of advanced scientific products for sale to industry, a dedication to working with customers to help them solve their problems, and a conscious avoidance of consumer goods and commodity products, continue to underlie the achievements of Rohm and Haas today, much as they have for seventy-five years.

PART ONE ·

SUPPLIER TO THE LEATHER AND TEXTILE TRADES 1909 ⋈ 1933

CHAPTER 1

THE BEGINNINGS
OF ROHM AND HAAS

America as the twentieth century dawned: a land of legend, a land of opportunity for the people of Europe, a land where a man of modest beginnings might make his fortune. Millions of Europeans streamed through ports of entry every year. In 1901 Otto Haas of Stuttgart, Germany, was one such arrival.

The twenty-nine-year-old Haas was in some ways a typical immigrant. His motivation was a common one: he believed that in America, if he worked hard, he could lift himself to a better life than would have been possible in his home country. His family was impoverished and had been unable to support his education past the age of twelve. From fifteen, he had been the primary support of his mother and younger siblings. But in many ways, Haas was more fortunate than other new arrivals. His family was considered middle class despite its poverty, having fallen on hard times only after the early death of Haas's father, a cavalryman turned civil servant. The family's position earned Haas a place as an apprentice in a bank. The ability he demonstrated there eventually gained for him a position as a clerk with responsibility for the firm's English accounts, in which role he learned to read and speak English. Eventually, he left the bank for a clerkship with G. Siegle and Co., a dye and chemical manufacturer, where he took part in the operation of a technically sophisticated business. The German dyestuffs industry, which the young Haas joined, was in many ways the prototype for the entire modern chemical industry. It was grounded in and continuously supported scientific research; its products were advanced synthetic chemicals for sale to other manufacturers, and it sold its products through technically trained salesmen who could work to solve customers' problems. When Haas arrived in America, he already possessed skills and had a career in which he had made significant progress.

Haas brought one other thing with him to the United States: an appointment to a position in Siegle's branch office on Staten Island, and thus the certainty of employment. He stayed with Siegle for two years before taking a position with the New York exporting department of Sulzberger and Company, one of the five Chicago-based companies that dominated the meat-packing business. The Sulzberger company management liked Haas and promoted him several times, but could not provide him with the opportunity he craved. What Haas wanted was a business of his own.[1]

Meanwhile, back in Germany, his friend Otto Röhm nurtured similar hopes. Röhm, four years younger than Haas, had trained as an apothecary's apprentice before turning to chemistry, studying first at the University of Munich, and then at the University of Tübingen. He was graduated with a Ph.D. in 1901, after completing a dissertation on acrylic acid and its derivatives under the direction of Hans von Pechmann. While at Tübingen, Röhm also came under the influence of Eduard Büchner, one of the pioneers in enzyme chemistry.

After several short-lived positions, Röhm obtained employment as an analytical chemist at the Stuttgart Municipal Gas Works in 1904. He soon concluded, however, that a career as a municipal chemist was too modest a goal.

Along with his official duties, he undertook private research, which he hoped would lead to a scientific discovery with commercial potential. Röhm had ample reason to believe that chemistry could lead him to prominence. German chemistry was the most highly advanced in the world, and the discipline contained many practitioners who had gained fame and wealth through practical application of their knowledge.

An odor emanating from a tannery near the Stuttgart Gas Works inadvertently launched Röhm's new career as a scientist–entrepreneur and led to his partnership with Otto Haas. Turning animal hides into leather is one of the oldest of industries, but procedures used in 1904 differed little from those of centuries past. First the hides were dehaired through soaking in a lime (calcium hydroxide) solution. Then came the "bating" or softening of the dehaired hides when they were placed in a solution of fermented dog dung. The odor from this process was terrible, and the results were inconsistent. Fermented dog manure varied in composition, and this variation affected the amount of soaking needed to soften the hides. It was impossible to predict how long the process would take; an artisan had to judge progress by feeling the excrement-soaked hides.

Röhm, sniffing the tannery odors, noticed that the tannery smell was similar to that given off by the gas work's waste product—gas water. He knew that gas water contained ammonia, carbon dioxide, and hydrogen sulfide. He wondered if a more efficient bating solution might be developed from the chemicals in the gas water. Röhm enlisted the cooperation of a local tanner who provided him with a place to perform the necessary experiments. By May 1906 Röhm had developed a composition, based on concentrated gas water and a few simple salts, which he believed would work. Several batches of goat skins treated with his invention appeared successfully bated. Röhm interested several tanners in the Stuttgart area in his compound and signed a development and manufacturing agreement with one of them. Thus encouraged, he resigned his position at the gas works. He wrote to his friend Haas, telling of his success and asking Haas to investigate the American tanning industry's consumption of dog manure (which proved to be substantial). He later invited Haas to return to Germany to form a partnership which would exploit the new bate. After further correspondence, Haas agreed to Röhm's proposition, but only on the condition that, once the business was established in Germany, he would return to America and run an American branch. With Haas's agreement, Röhm named his bate Oroh, from the two men's initials.

It took Haas until December to wind up his affairs in New York and return to Germany. He arrived to bad news. During the intervening months, Röhm had encountered problems that convinced him that Oroh did not work reliably. Instead of going out and selling a product, Haas became Röhm's unpaid laboratory assistant.

The failure, rather than discouraging Röhm, set him to studying the controversial bating process. He found that while the English leather chemist J. T. Wood believed bacteria in the dung solution caused the bating, others contended it was caused by the reaction of the leftover lime with a component of the bate. Röhm concluded that both explanations were partly right. In an effective bating operation, two different processes occurred, often simultaneously. Reaction with an acid or an acid salt removed the excess lime, while other components in the bate softened the hide. Röhm realized in May 1907, when he successfully bated a hide by treating it first with dog dung and then with Oroh, that his product accomplished only the deliming. What Röhm had learned from Eduard Büchner at Tübingen now enabled him to reach the crucial conclusion that had eluded Wood and all others. Büchner had challenged Pasteur's theory that bacterial fermentation was an inherently physiological process occurring only as an expression of the life force of the bacteria. Büchner found that fermentation results from ordinary chemical reactions of compounds called enzymes, which are produced by living cells. Enzyme action is independent of the producing organism. For his work, Büchner received the Nobel Prize in chemistry in 1907.[2]

Röhm was the first to realize that bating was an enzymatic process. The softening action of a bate was the result of enzyme-catalyzed decomposition of a part of the hide. Previous attempts to produce an improved bate had failed. The products were either, like Oroh, mere deliming agents, or they relied on bacteria, whose rates of enzyme production were easily disturbed by minor changes in reaction conditions. What Röhm now needed was an enzyme that would yield the desired bating effect. He quickly dismissed isolation of the enzyme from the manure solution; it would be too expensive, and there would likely be many other enzymes present. It seemed preferable to find a glandular enzyme with the proper activity. Within a few weeks he discovered that the aqueous extract of animal pancreases possessed the requisite property. In June 1907 Röhm filed a German patent application for a process of bating hides using the aqueous extract of the pancreas with the addition of appropriate mineral salts. He registered the trademark Oropon for his discovery. He derived this name from his initials and *opas*, the Greek word for juice.

Over the next several months, Röhm developed Oropon to the commercial stage. Realizing that standardization of the product would be one of its major selling points, he developed an analytical technique for measuring the enzymatic content of batches of pancreas juice. This enabled him to sell Oropon at standard strengths. The commercial Oropon would thus be an aqueous solution of pancreas extract to which ammonium chloride, a deliming salt, had been added.

Now that they finally were ready to begin, Röhm and Haas formalized their business arrangement by drafting, signing, and registering a partnership agreement on August 30, 1907. Röhm, for his share, committed his current and future inventions. Haas, for his, invested ten thousand marks which he had

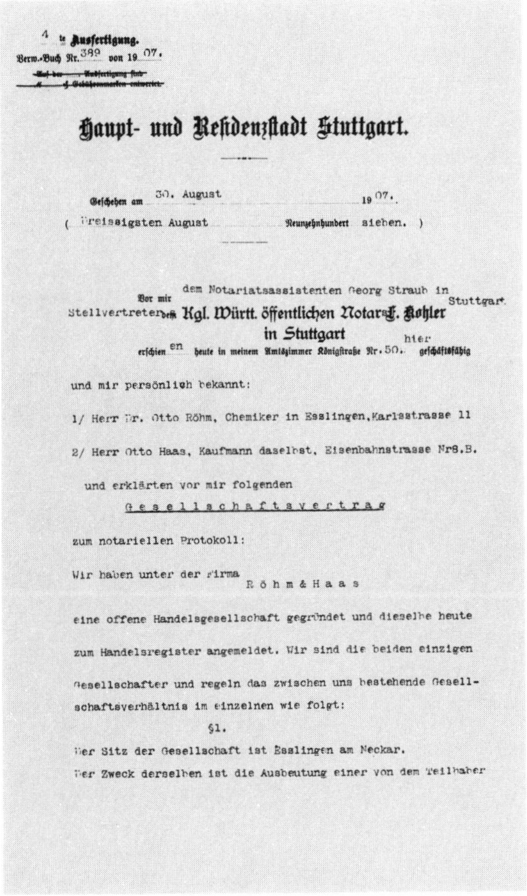

4 te Ausfertigung.
Verw.-Buch Nr. 389 von 19 07.
~~Auf der ... Ausfertigung sind~~
~~ℳ ... ₰ Gebührenmarken entwertet.~~

Haupt- und Residenzstadt Stuttgart.

Geschehen am 30. August 19 07.
(Dreissigsten August Neunzehnhundert sieben.)

Vor mir dem Notariatsassistenten Georg Straub in Stuttgart, Stellvertreter des Kgl. Württ. öffentlichen Notars F. Kohler in Stuttgart erschienen heute in meinem Amtszimmer Königstraße Nr. 50. hier geschäftsfähig und mir persönlich bekannt:

1/ Herr Dr. Otto Röhm, Chemiker in Esslingen, Karlsstrasse 11

2/ Herr Otto Haas, Kaufmann daselbst, Eisenbahnstrasse Nr8.B.

und erklärten vor mir folgenden

G e s e l l s c h a f t s v e r t r a g

zum notariellen Protokoll:

Wir haben unter der Firma

R ö h m & H a a s

eine offene Handelsgesellschaft gegründet und dieselbe heute zum Handelsregister angemeldet. Wir sind die beiden einzigen Gesellschafter und regeln das zwischen uns bestehende Gesellschaftsverhältnis im einzelnen wie folgt:

§1.

Der Sitz der Gesellschaft ist Esslingen am Neckar.
Der Zweck derselben ist die Ausbeutung einer von dem Teilhaber

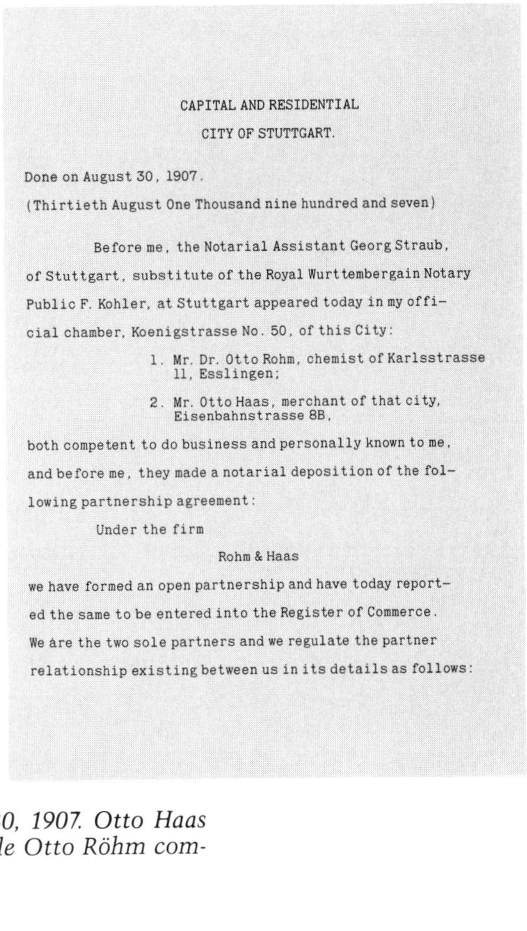

CAPITAL AND RESIDENTIAL
CITY OF STUTTGART.

Done on August 30, 1907.
(Thirtieth August One Thousand nine hundred and seven)

Before me, the Notarial Assistant Georg Straub, of Stuttgart, substitute of the Royal Wurttembergain Notary Public F. Kohler, at Stuttgart appeared today in my official chamber, Koenigstrasse No. 50, of this City:

1. Mr. Dr. Otto Rohm, chemist of Karlsstrasse 11, Esslingen;

2. Mr. Otto Haas, merchant of that city, Eisenbahnstrasse 8B,

both competent to do business and personally known to me, and before me, they made a notarial deposition of the following partnership agreement:

Under the firm

Rohm & Haas

we have formed an open partnership and have today reported the same to be entered into the Register of Commerce. We are the two sole partners and we regulate the partner relationship existing between us in its details as follows:

The partnership papers were signed on August 30, 1907. Otto Haas invested ten thousand marks in the venture, while Otto Röhm committed his present and future inventions.

Röhm and Haas's first manufacturing site was in Esslingen, just outside the city of Stuttgart.

acquired from a nobleman who had befriended him during his banking days. The new partnership established its first plant in rented quarters in Esslingen, a small city just outside Stuttgart.

The manufacturing facilities were modest, and the two partners did most of the work themselves, first putting the pancreases through a meat grinder and then squeezing the ground organs in an old manual press. They soon accumulated a basement full of rotting pancreas meat, which Haas disposed of (or so at least the story is told) by tossing it, one chunk at a time, off a bridge over the Neckar River on late-night walks.[3] Soon the partners hired workmen to handle the actual grinding and pressing. Within their first year of operation, they gained the trade of at least forty tanneries.

Haas may have had no technical training, but he knew from his years with a dye manufacturer how a technically advanced chemical could be sold to industry, and during the months with

Röhm he had learned practical leather chemistry. This knowledge enabled Haas, as well as Röhm, to spend much of the next two years on the road visiting tanners and convincing these skeptical businessmen and artisans through lengthy demonstrations that, unlike previously promoted modern bates, Oropon really worked. Haas discovered that tanners of different types of leather required different properties in their bates. Preparation of soft leathers, such as used in kid gloves, required strong enzymatic action, while tough leathers, for uses like soles and industrial belting, mainly required deliming. In response to this discovery, Röhm and Haas soon marketed at least a half-dozen grades of Oropon which differed in their relative proportions of enzyme and acid-salt delimer. The new bate, however, achieved its greatest market penetration among the soft leather tanners.

In the summer of 1908 the partners encountered a problem with their product: it tended to decompose if stored long in hot weather. This decomposition yielded a foul-smelling liquid which had neither of the advantages that had made Oropon a commercial success—standardized reproducibility and hygiene. Röhm first tried preservatives, but they did not work. He concluded that it was probably impossible to maintain enzyme stability in aqueous solution. What was needed was another medium to carry the enzymes. After trying several possibilities, he developed a process for adsorbing the juice onto powdered sawdust and then drying the product. This dry product proved to be stable. Röhm and Haas converted their production to this new form in the spring of 1909 and phased out the liquid bate. From then on, Oropon would be a mixture of pancreatic extract of standardized strength adsorbed on wood flour and mixed with an acid-salt deliming agent. One packet of the new dry Oropon, dissolved in water, equaled one kilogram of liquid Oropon in activity.

By 1908 so many German tanners were using Oropon that Haas traveled to England and France to introduce the product there. And by the following year French sales were sufficient to warrant the opening of an office in Lyons under the direction of Röhm's brother, Adolf. Due to the product's advantages and the men's salesmanship, Oropon became a standard item of commerce in Europe.

In the summer of 1909, the partners concluded that their growing business needed more space for manufacturing. After some searching, Röhm found a suitable vacant factory in Darmstadt, a medium-sized industrial city approximately eighty miles north of Esslingen. They moved there, and the German company has remained in Darmstadt ever since.

With the business thriving, Otto Haas made plans to return to America. Although he had lived in New York during his previous sojourn, he decided to settle in Philadelphia. What prompted this important decision was that the Philadelphia area was a tanning center, especially for the manufacture of leather for kid gloves. In June and July Haas shipped several casks of Oropon to America; he himself followed in August. On September 1, 1909, the American branch of Röhm and Haas opened for business in rented quarters at 202 North Second Street, Philadelphia.* At the age of thirty-seven, Haas finally was launching the enterprise that would make his fortune. By combining the greater opportunities available in the open and expanding American economy with the superior innovations inherent in world-leading German chemical technology, he expected to have the best of both worlds.

From his last two years in Germany, Haas knew much of what to expect. The tanners would have to be courted, provided with samples, and convinced through lengthy demonstrations that it was in their interest to use Oropon. The tenacity this selling required can be gleaned from a few pages of Rohm and Haas loose-leaf customer notes that have survived the years. These concern the firm of Robert H. Foerderer Inc. of Philadelphia. Haas's initial entry noted that Foerderer manufactured glazed kid, using dog manure as a bate. In 1909 Foerderer purchased small quantities of Oropon, type A (a grade containing one part pancreas extract adsorbed on wood flour to two parts ammonium chloride), but found it unsatisfactory: the leather was "too loose." In early 1910 Haas sent a sample of type D (a one-to-one mixture), but it apparently was not used, as the tanner was looking for a new superintendent. The ledger noted the appearance of two new superintendents within the year. Haas interested the first, but could not obtain Mr. Foerderer's permission to experiment. The second, a Mr. Best, "does not wish to make any changes for the present as his hands are full with other troubles."[4]

The next entry was dated April 5, 1912. It reported that "when Mr. Foerderer returns, we will call upon him for permission to experiment."[5] The tanner finally acquiesced; in June Haas sent him five-pound samples of two different grades and

---

*Hereafter, the umlaut will be omitted from the name of the American Rohm and Haas operation, although it was used by the firm until 1943.

instructions for their use. Best prepared several test batches of hides but was dissatisfied with the grain of the finished leather. He agreed that Oropon might be useful for other types of hide. There the matter stood until the following March, when Haas persuaded Best and Foerderer to allow a salesman to run an experiment-demonstration in the tannery with Oropon AB, a new type containing boric acid as an additional delimer. This experiment consisted of three packs of hides handled under differing conditions of time, temperature, and concentration, alongside a control batch using dog manure. The trial impressed Best; in April he purchased 150 pounds of Oropon AB, enough to bate 15,000 pounds of hides. The packs Best processed with this purchase proved so satisfactory that in late May he decided to switch production to Oropon; from that

In 1912 Rohm and Haas moved to larger quarters at 40 North Front Street in Philadelphia.

point on the ledger records purchases at least weekly, with monthly totals in the neighborhood of 2,000 pounds, or $700.

Through many similar successes, the Philadelphia office became profitable and expanded. Sales grew from $14,608 in 1910 to $47,981 in 1911, and the branch no longer needed financial subsidy from the Darmstadt office of the partnership. It even returned a profit of $7,530. In 1910, because of the amount of time he spent working in the tanneries, Haas hired an office manager, Carl Herbster. Haas's decision to delegate office responsibilities rather than sales is characteristic of the importance he placed on working with current or potential customers. Sales nearly doubled in 1912, to $89,542; Haas now had more customers than he could handle alone. Late in the year he hired an experienced leather chemicals salesman, a German immigrant named Hugo Schaefer. With these men, the company expanded its market to other parts of the East. The business also outgrew its original facility and moved to larger quarters at 40 North Front Street.

As other parts of the country had important tanneries, Haas made several western trips, concentrating on Chicago and Milwaukee. In February 1912 he opened a branch office in Chicago with its own full-time salesman, George Callender. A second salesman, Carl Muckenhirn, joined Callender in Chicago the following year; he would remain there into the 1950s. Both Callender, a Scot, and Muckenhirn, a native of Detroit, had graduated from the leather chemistry program at the University of Leeds in England, the finest such program of the era.[6] With these early employees, Haas instituted a policy which would guide the company for many years: hire as salesmen only those individuals whose technical training would enable them to understand and solve the customer's problems. These problems were becoming more varied, for the use of Oropon spread to tanners of other types of leather. By 1916 the product was in common use for every type except sole leather.

In the fall of 1913 his business was running so smoothly that Haas took a two-month trip to South America to establish Oropon there. Although German firms generally retained expansion rights for themselves, the Philadelphia office was not a subsidiary, but a branch run by a full partner. For this reason, Haas and Röhm had agreed earlier that South America could be served best from the United States. Even before Haas made his trip, small amounts of American-packaged Oropon had been sold through agents in several South American countries. Haas expanded the business by establishing branch offices,

Charles Hollander joined Rohm and Haas on February 16, 1914, as a
chemist in charge of the laboratory and plant in Chicago. In 1917 he
was appointed manager of the Bristol plant and head of research
there.

staffed with Rohm and Haas employees, in both Buenos Aires,
Argentina, and Santiago, Chile. The trip proved important to
Haas for personal reasons as well: on board ship, he met his
future wife, Dr. Phoebe Waterman, a University of California-
trained astronomer who was on her way to observe the southern sky.

American manufacturing evolved at a rate commensurate
with sales. Originally Haas imported the various grades of
Oropon from the Darmstadt plant. In 1911 he imported just the
adsorbed extract, which workmen in his back room mixed with
appropriate amounts of deliming salts to produce the several
grades. In 1914 Haas was ready to take the big plunge and open
a full factory. He chose Chicago because it was the center of
the meat-packing industry, the source for pancreases. He rented
a factory at 2252 South Western Avenue in February. Under

George Callender's general supervision, the plant started its first batch on June 23, 1914, and produced Oropon in commercial quantities beginning in July.

Callender died that summer after a brief illness. In his place Haas hired Dr. Charles Hollander, an American who, like most American Ph.D. chemists of the era, had earned a German degree (in this case under Nobel Laureate Richard Willstätter at the University of Munich). Röhm helped get the plant operating smoothly by temporarily assigning his assistant, Dr. Karl Stutz, who was intimately familiar with the Darmstadt plant, to Chicago. Many years later Muckenhirn recalled that Stutz was often less than helpful, an individual who would reply to disagreements by citing the authority of *"Die Herren in Darmstadt."*[7] Nonetheless, Stutz helped Haas become independent of Darmstadt for his supply of Oropon; the Chicago plant's production met the demands of its customers.

Haas was either lucky or perspicacious in the timing of the Chicago plant opening in that the outbreak of World War I in July 1914 shattered Europe's normal trade relations with the Western Hemisphere. The British naval blockade, German submarine warfare, and diversion of the industrial capacity of the combatants to war materiel production, all contributed to the disruption. For the American chemical industry, this was a time of confusion, shortages, and unparalleled opportunity. Through a combination of economic might and patent control, a group of German companies controlled the worldwide fine chemical market, especially in technologically advanced products such as dyes and medicines synthesized from coal tar. These chemicals all but disappeared from America; prices increased tenfold and more for the little material available. The United States had greater self-sufficiency in many of the more basic heavy chemicals, but even there shortages occurred, as some raw materials, including potash and nitrate, were imported.[8] Haas's position was, by contrast, very strong; the Chicago plant enabled him to continue his business unimpaired.

Haas was shrewd enough to see an opportunity for expansion in the midst of the confusion. By this time he knew all the chemicals used in the tanneries, including many that American manufacturers produced in ample supply. He also had become familiar with the chemicals involved in textile manufacture. The branches set up the previous year in Chile and Argentina now provided attractive outlets through which Haas could supply these items to the South American market as the disruption of established international trade patterns

increased. In South America, the chemical industry was rudi-
mentary at best, so there had been a greater dependence on
European, and chiefly German, imports. Haas quickly put together an extremely profitable export division, which, although
it did sell Oropon, mainly served as a jobber for leather and
textile chemicals such as sodium sulfide, a dehairing agent.
Haas purchased these chemicals domestically and sold them
through his agents in South America. Haas's background in the
German chemical industry again served him and his salesmen
well. They knew the German style of doing business and thus
could smoothly replace the German suppliers. During the next
three years, this export division provided Rohm and Haas with
a profit equal to one-half of its profit from domestic sales.

The relationship Haas had established with tanners and
other chemical companies, along with his new exporting experience, led him to attempt a diversification of his domestic
manufacture. By 1917 Rohm and Haas manufactured and sold
modest quantities of several other tannery chemicals, including a one-bath chrome tan, leather finishes, fat-liquors, and
Titanene, a mordant for dyeing. Chemically, this last product
was titanium sodium sulphate, a wartime replacement for the
scarce titanium potassium oxalate.

This diversification contributed to Haas's growing dissatis-
faction with the location of his plant. As the company grew
and added more products, the need for a facility larger than the
one in Chicago became apparent. Moreover, Haas could not
maintain the close personal supervision he preferred from an
office a thousand miles away. Nor did he consider it desirable
to move his office to Chicago; his largest customers remained
in the East and he had become settled and established in the
Philadelphia business community. He concluded that he would
rather ship pancreases than Oropon east from Chicago.

Haas decided to relocate the plant on land with room for
expansion in the vicinity of Philadelphia. He and his wife spent
many Sunday afternoons in 1916 taking local trains to different
parts of the metropolitan area, and then scouting for possible
plant sites.[9] In November 1916 he settled on fifteen acres of
farmland in Bristol, Pennsylvania, some twenty miles upriver
from Philadelphia. The partnership purchased the tract on December 5, 1916. It boasted a rail line (the New York Division of
the Pennsylvania Railroad), as well as access to the Delaware
River. The following summer, Otto Haas personally purchased
an adjoining plot of almost forty-five acres.[10]

Groundbreaking for the initial group of buildings occurred

By September 1917 construction of the Bristol plant was well under way. The first batch of Oropon was made here on December 17 of that year.

The original construction at Bristol included two manufacturing buildings, a power house, a garage, a storage building, a bungalow, and a building for the analytical lab and plant office.

early in 1917, but winter weather prevented substantial progress before April. It took most of the remainder of the year to complete construction and equip the buildings. The Bristol plant produced its first batch of Oropon on December 17, 1917. The batch came out of Building 1, the same Oropon building that stands today. As might be expected, Building 1, the site for production of the company's main product, was the largest and most elaborate of the seven buildings in the original complex. It contained four stories and 21,300 square feet of floor space. An appraisal made the following year valued the building at $64,000 and the equipment at an additional $11,400. The original construction also included four other buildings, and several small sheds. One of these buildings housed the plant office and the analytical laboratory where the Oropon would be standardized. The 1918 appraisal valued the entire plant, including land, buildings, and equipment, at $287,000, which represented a remarkable success story for an operation that had begun from scratch less than a decade before.[11]

Otto Haas's gate pass

Oropon was shipped from the Bristol plant in barrels made in the same production building.

Otto Haas hired Stanton Kelton, Sr., a Harvard-trained lawyer, in 1916, because he thought that a man with an orderly legal mind could prove his worth through organizational talents.

During the first half of 1917, the Chicago plant accelerated its production of Oropon to carry Rohm and Haas through the interval before the Bristol operation would open. The Chicago plant manufactured its last batch on July 16, 1917, after which the employees dismantled the equipment. Haas then gave them all the option of relocating in Pennsylvania, except for Muckenhirn who remained as the western salesman.

The company continued to grow, achieving a profit in 1916 of $212,213 on sales of $806,450. This growth required a continual expansion of staff. Among the people Haas hired was a

Harvard-trained lawyer, Stanton Kelton, Sr. Haas didn't require a full-time in-house counsel, but he believed that a man with an orderly legal mind could prove his worth through organizational talents.

If Haas had taken the time, in the spring of 1917, to sit and reflect upon what he had accomplished since accepting Röhm's invitation some ten years earlier, he surely would have been pleased. Starting with little besides his own wits and his friend Röhm's scientific talents, he had built a very profitable business, the sales of which were approaching the million dollar mark. He had achieved this early success, it was true, in a decade when the American industrial economy was prosperous and expanding, and receptive to new ideas. But even in those conditions, far more new ventures failed than succeeded. Haas could attribute his success to his own skills as a businessman: personal attention to technical sales, careful control of costs and capital, insistence on maintaining quality control. If he had taken further time for thought, he would have recognized much of his successful technique had its roots in the things he had learned while working in the German dye industry. With his talents, background, personal drive, and entrepreneurial zeal, Haas himself, as well as his chief product, Oropon, had become well established within the industry. He had a large and modern plant under construction, which had been paid for out of retained earnings. He had found a wife and started a family. Here, surely, was a man who had reason to believe in the American Dream.

Haas may have been riding on a train of hard work and uninterrupted success, but external events were conspiring to lay an ambush. Haas was still a subject of the German Kaiser, as was his friend and partner, Röhm. On April 16, 1917, Woodrow Wilson signed a declaration of war against Germany. Neither Haas nor his business would ever be the same.

NOTES

This chapter is based on documents in the Rohm and Haas Company archives in Philadelphia, with the exception of material on Otto Röhm and the Röhm and Haas Company in Germany, which comes from Ernst Trommsdorff, Otto Röhm: Chemiker und Unternehmer *(Düsseldorf: Econ Verlag, 1976).*
    1. Bill Kohler, interview with the author, July 1, 1983, copy in Rohm and Haas Company archives. Testimony prepared by Otto Haas for 1945 antitrust trial.

2. Herbert Schriefers, "Eduard Büchner," The Dictionary of Scientific Biography, ed. Charles Gillispie (New York: C. Scribner & Sons, 1970–80), 2:560–63.

3. Donald Frederick, interview with the author, September 27, 1983, copy in Rohm and Haas Company archives.

4. Loose-leaf customer records, file for Robert H. Foerderer Inc., Rohm and Haas Company archives.

5. Foerderer file.

6. Letter, Carl Muckenhirn to Phillip Lind, January 14, 1959, Rohm and Haas Company archives. This is a six-page reminiscence Muckenhirn wrote at Lind's request.

7. Muckenhirn letter.

8. Williams Haynes, The American Chemical Industry: A History, 6 vols. (New York, D. Van Nostrand, 1945–54), 3:3–54.

9. Kohler interview.

10. Bound Minute Books of Rohm and Haas Company, Board of Directors Meeting, May 20, 1920, 1:114.

11. Morris Knowles, Inc., The Report upon Rohm and Haas Company, November 19, 1918, records of the Office of the Alien Property Custodian, Corporate Management Division, Record Group 131, File CM 1804, National Archives, Washington D.C. (hereafter APC).

WORLD WAR I AND AFTER ⊢ AN INDEPENDENT AMERICAN COMPANY

For the first time in history, the United States found itself fighting in Europe. Germany's wide-ranging submarines, which attacked both American merchant vessels and passenger lines, did much to end United States neutrality. The British acted little better; their ships intercepted American vessels destined for neutral ports and searched for goods intended for Germany and its partners. But the British, who controlled the seas, were able to do this without loss of life and avoided provoking an American response.

The declaration of war gave Haas concern for the future of his hard-won business. Anti-German feelings ran high throughout the United States. The Kaiser was hated, and German soldiers were characterized as evil. Loyal German-Americans faced torrents of abuse from their fellow citizens. As a German subject running an American company that was vital to the war effort, and with a partner in Germany presumably aiding the enemy, Haas was in an extremely vulnerable position.

His outside counsel, Douglas Moffat of the New York law firm Cravath and Henderson, convinced Haas the best way to safeguard his rights under wartime constraints was to incorporate the business. However, distribution of stock for the new corporation was no simple matter. The main books of the partnership which might demonstrate the interests of the partners were in Germany with Röhm and inaccessible. The partnership agreement was of little help, as it provided one formula for distribution of profits and another for distribution of losses, and existed alongside an agreement that Haas would have managerial control over the American house. In any case, legal opinion held that the declaration of war dissolved all partnerships such as Rohm and Haas. Moffat advised Haas simply to hold all the stock himself as trustee for "the interests of the partnership as they might appear." Moving swiftly, Haas had Rohm and Haas Company incorporated in Delaware on April 24, 1917, just eighteen days after the United States entered the war. The new company obtained the American, South American, and Spanish assets of the old partnership in return for $600,000 worth of stock, the entire amount issued.[1]

Haas had been shipping Oropon to Britain since 1915 (after convincing the British his was an American business and not a German front). He also had shipped it to South America, using the name of a friend, Patrick Doheny, on some shipments handled by firms that might not have dealt with a firm with a German name.

With America's entry into the war, demand for Oropon increased. It was used to bate leather for the upper sections of Allied combat boots. Despite its small size and limited product line, Rohm and Haas Company played an important role in the manufacture of a mundane but essential piece of military gear.

During this period United States authorities had good reason to monitor businesses with German roots or ownership. Some of them secretly had been aiding the German war effort. For example, the American subsidiaries of two German chemical manufacturers, Heyden Chemical Works and the Bayer Company, conspired with an American firm to corner the market in phenol and convert the chemical into aspirin, making it unavailable for explosive manufacture. In response to these and other concerns, Congress passed the Trading with the Enemy Act on October 6, 1917. This act established the Office of the Alien Property Custodian (APC) with authority to seize alien-owned assets. The APC quickly moved to seize the American branches of firms like Heyden and Bayer and to replace their managements with American citizens.[2]

The Rohm and Haas Company soon came to the attention of Alien Property Custodian A. Mitchell Palmer. Haas submitted a long affidavit in which he described Röhm's interest, but asserted Rohm and Haas of Philadelphia always had been run separately and independently of Darmstadt. He presented a history of the company designed to support his claim and explained the incorporation as a measure taken to expedite foreign shipments, improve tax status, and "partly to avoid the technical embarrassment of having a German partner." The War Trade Board sent the APC a copy of its file on Rohm and

Haas, adding that the enemy ownership was "clear and admit-
ted" and pointing out with disapproval that Rohm and Haas
had used other names on foreign shipments. The Board urged
that the enemy interest be seized promptly.[3]

Haas won the support of the Tanners Council of the United
States of America, an umbrella group formed by three leather
trade associations to coordinate the industry's part in the war
effort. The Tanners Council membership included virtually all
of the important tanners and leather finishers. Its representatives wrote to the APC on Haas's behalf, calling Oropon essential to their industry and seeking assurances the government
would do nothing to interfere with Oropon production. They
also arranged for Haas to meet with an APC representative to
plead his own case.

Haas's appeal was successful; he convinced the government
that although he still held German citizenship, his loyalties
were with his adopted land. His interest in and management of
Rohm and Haas could therefore continue undisturbed, at least
for the time being. The APC agreed that the government's interest should be restricted to Dr. Röhm's share. It simply required Haas, on December 31, 1917, to name a local
businessman, W. W. Washburne, as the APC's representative on
the company's five-man board of directors. Washburne then
chose an overseer whose job it was to see that the company's
goods and funds were handled in an appropriately patriotic
manner. Haas perceived these changes as government certifi-
cation of his firm's American credentials.

Washburne's directorship proved beneficial to Rohm and
Haas. With his assistance, Haas convinced the War Trade Board
to grant Corporate Secretary Stanton Kelton permission to go
to South America to straighten out the firm's branches there.
Kelton appointed new agents acceptable to the local American
consuls. After that, Rohm and Haas's export activities, which
had been halted by the War Trade Board, were allowed to re-
sume.

Although it kept Rohm and Haas under investigation, the
APC took no further action for a year. It did not seize Dr.
Röhm's share of the company, in large part because no one
could determine what that share was. Furthermore, the APC
questioned the legality of Rohm and Haas's incorporation. The
confusion led to extended negotiations involving Haas, Washburne, the Tanners Council, the APC, and attorneys for all.
The APC attorney recommended the incorporation be held
illegal and the company be thrown back into a partnership. The

APC could then either liquidate the partnership or sell it. Under this scenario, Haas, as a German national, would have been barred from reacquiring his own company. Palmer rejected the recommendation. The Tanners Council had persuaded him Oropon was essential in the tanneries and an independent investigator Palmer had hired convinced him the key to Oropon manufacture and use lay in know-how beyond that disclosed in the patents, and the company and product had no value without Otto Haas. The APC therefore could act within the spirit of the law (which was that alien-held property be managed in the best interest of the Allied war effort) only by ratifying the incorporation.

The question then became one of determining, in the absence of appropriate records, what was the "alien interest" that the APC should seize. The APC first demanded sixty percent, the percentage of the profits to which Röhm was entitled under the partnership agreement, but Haas insisted that he would not continue with the company unless he retained control. As Moffat wrote to the APC, Haas's desire to be his own boss was a good part of the reason he had come to America in the first place. Once again, the tanners intervened on Haas's behalf. In Chicago, the Tanners Council had set up a cooperatively owned, for-profit affiliate known as the Tanners Products Company. Tanners Products, through its president, Van Wallin, offered to purchase the alien interest in Rohm and Haas from the APC regardless of its final form. Wallin would accept minority interest or joint control, if that was necessary to keep Haas in the business of supplying Oropon for the tanners. This arrangement suited Haas. He trusted the Tanners Council leaders and believed he could work with them, even if he had to yield majority ownership.

Initially, Palmer accepted the Tanners Proposal, and all parties agreed on a plan. The APC would seize fifty percent of the outstanding Rohm and Haas stock as Röhm's interest. Haas would deposit an additional ten percent of the stock with a trust company, which would hold it pending a postwar determination of the partners' interests. The APC would then sell at auction both its fifty percent and an option on any of the ten percent that might come into its possession. Tanners Products would presumably be the sole bidder. Even though the war was by this time nearing its end, the APC went ahead with the plan. It scheduled a public auction of Röhm's interests in Rohm and Haas for December 14, 1918. On November 13, two days after the armistice, advertisements appeared in major newspapers, offering Röhm's unspecified interest in the partnership, the corporation, and/or the patents. The vague wording may have been an attempt to ensure that there would be no competing bidder. At any rate, the APC canceled the auction at the last minute because a suit filed in an unrelated case challenged its authority to make this type of sale. The end of the war also may have contributed to the cancellation.

Much to the annoyance of Haas and Wallin, the sale was not rescheduled for more than a year. Many other APC sales were similarly delayed, chiefly due to the constant postwar turnover of APC personnel. The newspaper announcements which appeared in February 1920 specified exactly what would be sold: three thousand shares of Rohm and Haas stock (or fifty percent of the total), an option on another six hundred shares, and Röhm's interests in certain patents. When the sale was finally held, on March 4, 1920, Tanners Products was the sole bidder and paid the government the prearranged sum of $300,000.

Later that year Haas traveled to Germany at APC's request to gain Röhm's acquiescence. The two men found the wartime separation had done little to change their friendship. Röhm signed an agreement ratifying the incorporation and recognizing that his share in the United States business had been no more than half. Haas signed an agreement which stated that his share of the German business was forty percent, and the Darmstadt operation was incorporated on that basis as Röhm and Haas A.G. Haas further agreed to attempt to reacquire Röhm's share of the American company from Tanners Products so the scientist's interest could be reestablished.[4] These agreements were signed without reference to the German partnership books, which Haas had claimed since 1917 would show the true apportionment of the businesses. The APC accepted Röhm's signature as sufficient to allow the release of the remaining six hundred shares of Rohm and Haas stock to Haas in 1921, thus giving him fifty percent ownership.

After the March 1920 sale, three representatives of Tanners Products joined the Rohm and Haas board, and Wallin became vice president and treasurer. As promised, Tanners Products left Haas in control, but required that weekly sales reports be provided and that Wallin approve any major moves. Wallin also stipulated that Haas prepare an annual report for Rohm and Haas and attend the annual meeting of Tanners Products. In practice Wallin seems to have agreed to the things Haas (who after all had proven himself to be a skillful businessman) wanted to do, and the relationship worked well.

Despite the extended legal maneuvering and the partial change of ownership, Rohm and Haas Company continued to grow. Haas, ever on the lookout for suitable new products, accepted the wartime challenge presented to him by the General Dyestuffs Corporation to become the first American producer of sodium hydrosulfite, an important reducing agent used to affix dyes to textiles. This was a tricky chemical to produce but in late 1918 Dr. Hollander succeeded in developing a satisfactory process.[5] In 1919 Rohm and Haas began selling the chemical under the trade name Lykopon. Soon the company introduced two related products: sodium sulfoxylate formaldehyde which it sold as Formopon, and zinc sulfoxylate formaldehyde which it sold as Formopon Extra. These reducing agents became major products which in 1923 surpassed Oropon in sales. They also established Rohm and Haas as a supplier of textile chemicals. In 1920 Haas introduced the dehairing agent Arazym, a new enzymatic leather preparation which he had obtained from Röhm before the war. Although surviving records do not explain the delay in its introduction, Haas may have learned of improvements to this product during his postwar visit with Röhm.

In January 1920 Haas learned that one of his suppliers, Charles Lennig and Co., a Philadelphia manufacturer of basic heavy chemicals, was for sale. Lennig had supplied Haas with the sodium sulfide he sold to South American tanners and also with sulfuric acid. Initially, Haas planned to bid on Lennig personally, perhaps to give himself a new business in the event someone other than Tanners Products purchased the APC-held Rohm and Haas stock.[6] After Tanners became co-owners, Rohm and Haas Company completed the Lennig purchase. It bought all but a handful of Lennig's stock for just under $1 million, two-thirds immediately and the remainder in 1922. Rohm and Haas then operated Lennig as a subsidiary rather than as a department in the parent company.

What Rohm and Haas got for its money was one of the oldest chemical manufacturers in the United States. Nicholas Lennig had founded the business in Philadelphia in 1819 to import French and German chemicals and drugs, but soon moved into manufacturing. In 1832 he erected a plant in the Port Richmond section of the city and became the city's first large-scale producer of sulfuric acid. A decade later his sons and successors purchased a new manufacturing site several miles upriver in what would become the Bridesburg section of Philadelphia. In 1847 they moved their plant to this new location. The Bridesburg plant has been in operation ever since, making it one of the oldest continuously operated chemical plants in the country. The Lennig Company business declined in the twentieth century. Indeed, it might well have gone out of business before 1920 except for the chemical industry boom that followed the outbreak of the European war. The plant buildings Rohm and Haas acquired were constructed mostly of wood and were in poor condition. The old firm employed only two chemists, one

The Tacony Chemical Works is just one of the names used by the Lennig Company. This office was at 112 South Front Street.

of whom was the plant superintendent. Lennig's most valuable assets were the plant site itself and a customer list which made it the largest supplier of heavy chemicals to Delaware Valley industries. Foremost among the customers was the Roebling Company, the famous builder of the Brooklyn Bridge, which used Lennig's sulfuric acid at its plant near Trenton in the manufacture of steel bridge cable.[7]

Haas had ambitious plans for Lennig. He directed the plant superintendent, Dr. Siegfried Kohn, a chemist, to keep Bridesburg's processes up to date and operating smoothly and to investigate (as time allowed) possible new products. He vowed to improve employee relations by providing better working conditions and decent wages. He planned to repair the decaying buildings (at what proved to be an expense of over $80,000 within the first year) and, where needed, to replace them with more modern structures of metal and brick.

Many of Haas's ideas for improving Lennig's operations were soon put on temporary hold. Rohm and Haas could not pay cash for Lennig, as Haas would have preferred, so it borrowed half of the funds on short-term notes, confident increased earnings would enable the sum to be paid back quickly. However, a major business slowdown made this impossible. Sales during the latter half of 1920 were at best half of the previous year's. Days passed when neither Rohm and Haas nor Lennig received any orders. These financial problems worried Haas; he did not like to be at the mercy of bankers. Tanners Products came to his assistance by negotiating a loan at a favorable rate with a Chicago bank. Tanners Products also agreed to accept the issuance of credit memoranda instead of cash dividends. When prosperity returned in the fall of 1921, Rohm and Haas began reducing its debt, but funds remained tight for another two years until Haas negotiated a long-term mortgage on the Bridesburg plant. He also began the modernization program he had planned for Bridesburg; by the end of the decade most of the buildings had been replaced. Haas's discomfort over this debt and financial crisis reinforced his intentions to maintain high liquidity and avoid borrowing in the future.

The four years ending in 1921 saw many changes for Rohm and Haas. It became an independent American corporation. Dr. Röhm lost his share of the business, and Tanners Products gained its. The firm survived the major threat of a government takeover and prospered during the war-induced boom in the American chemical industry. The success of Lykopon gave Rohm and Haas a second major line of manufacture—textile

These women (and two men) staffed the Lennig plant during World War I.

chemicals. With the purchase of Lennig, the company added a second plant, a source of supply of some of its raw materials and, through the subsidiary's sales department, a new and varied group of customers.

But Haas was not satisfied. He had a partner, Tanners Products Company of Chicago. Although they got along, he still had to answer to that partner. He felt his friend Röhm had been unjustly deprived of a share in the American business that had been founded on his genius. The business was profitable, but he wanted it to grow more rapidly. He wanted new products which, like Oropon, could be sold to manufacturers on the basis of unique technical advantages. Otto Haas was determined to attack these problems in the years to come, but without abandoning the strategies that had served him well to date: conservative financial planning, the production of advanced chemicals for use by other industries, and the sale of these goods by highly technically trained men who could work with their customers.

NOTES

Additional source material for this chapter comes from the Rohm and Haas file, CM 1804, in the records of the Office of the Alien Property Custodian (APC), Corporate Management Division, Record Group 131, National Archives, Washington, D.C.

1. *Affidavit of Otto Haas, November 17, 1917, APC.*

2. *Stanley Coben,* A. Mitchell Palmer: Politician *(New York: Columbia University Press, 1963), pp. 127–54. Haynes,* Chemical Industry, *2:124–33, 3:258–60.*

3. *Richard Ely, War Trade Board, to A. Mitchell Palmer, December 4, 1917, APC.*

4. *Agreements between Otto Röhm and Otto Haas, September 9, 1920, Rohm and Haas Company archives.*

5. *Frederick interview.*

6. *Douglas Moffat to H. J. Galloway, January 24, 1920, APC.*

7. *Stanton Kelton, "Outline of the History of Charles Lennig and Company, Incorporated, and Its Predecessors, 1819–1947," 1948, Rohm and Haas Company archives.*

CHAPTER 3

PROSPERITY AND DEPRESSION

The ever-industrious Otto Haas did far more than meet with Dr. Röhm on his 1920 European trip. He used his APC-sponsored travel permit to visit several large German chemical companies with the aim of learning whether new technologies they had developed during the war years might be employed profitably in America. The largest of the companies he visited, *Badische Anilin und Soda Fabrik*, was sufficiently impressed with Haas that it sent one of its chemist-executives, Dr. Carl Immerheiser, to Philadelphia the following April to present a business proposal concerning Ordoval, a synthetic tanning material, or syntan. Ordoval was one of many new products which the German chemical industry had developed in response to war-caused shortages of conventional materials.

Tanning, which immediately follows bating, is the central step in the manufacture of leather from hide. A tanning agent chemically reacts with the protein of the hide, producing a modified protein that will no longer readily decay or putrefy. Before Ordoval, two classes of tans were in common use: vegetable tans, such as sumac leaves and quebracho bark, and chrome tans, basic salts of the metal chromium. Rohm and Haas marketed a chrome tan under the name Koreon.

Badische had introduced Ordoval successfully in Europe, but could not sell it in the United States because the APC had seized its American subsidiaries. Immerheiser offered Haas the North American rights to manufacture and market the product, and a contract was signed in May 1921.[1] Haas renamed the product Sorbanol and gave Dr. Siegfried Kohn the task of adapting it to the American market. He decided that Lennig would manufacture the product in Bridesburg and sell it to the parent firm on a cost-plus basis.

Badische itself had brought the first syntan, Nordoval, to the American market in 1913 for use as a modifier of vegetable tanning liquors. Nordoval attracted substantial favorable attention from tanners, but the European war soon forced Badische to discontinue its exports. After the United States entered the war, the American companies were able to work under seized alien patents. Several firms, including Rohm and Haas, introduced their own syntans. But the American chemical industry was not as sophisticated as that of Germany, and the domestic imitations were almost uniformly inferior. They served mainly to drive syntans into disrepute as chemicals that, likely as not, contained free acids which would damage the hides.[2]

In many ways the marketing problems presented by Sorbanol were similar to those Haas faced earlier in the introduction of Oropon. He had to overcome the skepticism that unsatisfactory experiences with wartime syntans engendered among tanners. Moreover, while Badische's European introduction had been aided by wartime shortages of natural tans, American tanners of the 1920s faced no such shortages. Convincing the tanners that Rohm and Haas syntans had advantages in price, process, or performance over natural tans proved difficult. Sales of the syntan remained modest in the first years after its 1922 introduction.

Prospects brightened in 1924 when Haas acquired Leukanol, an improved Badische syntan. Leukanol imparted a far superior color to the finished leather and replaced Sorbanol in most of its uses. Still, general acceptance did not occur until Rohm and Haas showed that leather treated with Leukanol along with a conventional tanning agent tanned more quickly and had superior properties. Then sales of the syntans boomed, increasing from 1.9 million pounds and $115,000 in 1926 to 7.4 million pounds and $484,000 in 1929. As with Oropon, patient, technically oriented sales development paid off in the end; Leukanol became a standard part of the tanner's art.

Acquisition of the syntans set the pattern for Rohm and Haas growth. Every summer or two from 1920 until World War II, Haas visited Germany, scouting its chemical industry for products he could profitably adapt for sale in North America. Here his position astride the two cultures of the old and new worlds served him well. German industrialists grew to know, respect, and trust him. He spoke their language and understood their industry. Thus, Haas had a real advantage as he became a sought-out, preferred conduit for American exploitation of the fruits of some of the world's best industrial chemical research. Among the products Haas acquired in this way were Ortho-chrom leather finishes and Tamol dispersant. In a sense, Haas was merely broadening his transfer of German technology to the United States to include inventions other than those of his friend Otto Röhm.

Haas was not content to rely solely on his skill as a negoti-ator for the development of new products. As early as 1914, when he hired Charles Hollander, he intended that his com-pany would develop products of its own. By the early 1920s, Rohm and Haas employed two Ph.D. chemists, Hollander at Bristol and Kohn at Bridesburg. Each had a staff of three or four chemists with lesser degrees, and each performed sufficient research to publish an occasional paper in journals of leather or textile chemistry. Rather than pioneering research, these papers, such as Kohn's 1922 publication on comparisons of different types of tanning materials, were aids for the establish-ment and support of extant Rohm and Haas products.[3]

Hollander and Kohn had limited time for research. They managed their respective plants and faced day-to-day operating problems that precluded much new product development. As Kohn commented to Haas in the cover letter to his "Progress Report for 1922": "We hope you realize as we do, that our major efforts were spent upon trying to keep up production under adverse conditions."[4]

Haas hired his first full-time scientific researcher not to develop new products but to defend an existing product against critical assault. The product was Oropon. A prominent scien-tist claimed it didn't work. To rebut the charge, Haas employed a young Ph.D. leather chemist, Harold Turley, in October 1924. Before Turley could complete his investigation, the charges against Oropon were withdrawn. Turley subsequently devel-oped successful fungus-based (and later mold-based) grades of Oropon. He continued on to a productive career as a scientist at Rohm and Haas, retiring in 1963 after many years as head of leather research.[5]

In the months following Turley's employment, Haas ex-panded the research department and gave it responsibility for technical sales but not for either plant operations or produc-tion. Some of the new chemists worked on development of existing product lines while others engaged in pioneering re-search.

Dr. Charles Peet was hired to investigate a new area: syn-

Harold Turley heard about the opening for a chemist at Rohm and Haas just as he was finishing up his Ph.D. at the University of London. He became Rohm and Haas's first full-time research scientist in 1924.

thetic organic insecticide research. Hollander saw this field as ripe for a breakthrough, and recommended it to Haas. The insecticides of 1926 fell into two broad categories: simple inorganic salts, such as lead arsenate, and naturally occurring organic compounds, such as pyrethrum, rotenone, and nicotine, which could be extracted from plants. In little more than a year, Peet identified a class of compounds, thiocyanates, that were strongly and immediately toxic to insects. Working with Anthony Grady, an entomologist, he developed a simple test to measure the effectiveness of insecticides. A measured number of insects would be released into a standard chamber, followed by the test compound. The dead insects would then be counted and a "kill ratio" calculated. Many a young technician started his career at Rohm and Haas counting dead flies. This Peet–Grady test became the industry standard, accepted both by insecticide trade groups and the U.S. Bureau of Standards.

Peet had been looking for a crop insecticide, but the thiocyanates he and Dr. Leon Heckert developed proved most effective as replacements for pyrethrum in household fly sprays. Rohm and Haas named Peet's discovery Lethane, from the words lethal and *thanatos* (which is Greek for death). The company introduced Lethane commercially in 1929 and, in keeping with his principle that sales should be handled by people with appropriate technical training, Haas transferred Peet to sales development, where he worked until his death in 1935.

Lethane gained acceptance slowly through the 1930s, selling as much for cattle sprays as for household use. Rohm and Haas's advertising touted the insecticide as bringing the same advantages to its potential customers that Oropon had brought to tanners: "Lethane 384 appeals to manufacturers because it is standard. . . . A scientific synthetic product, it does not fluctuate in strength and quality."[6] Lethane became a major product only when the outbreak of World War II halted imports of pyrethrum from the Orient. It never was an entirely satisfactory product; problems with human irritability plagued it through all its formulations and variants.[7] Nonetheless, Lethane was important for several reasons: it was the first commercial discovery of Rohm and Haas Company research; it represented the beginning of the company's major business in agricultural chemicals; and it was scientifically significant as the first commercial synthetic organic insecticide anywhere.

Within a few years of Peet's success, Rohm and Haas expanded its laboratories for research into areas including high-pressure catalysis, enzymology, and organic electrolysis. Using techniques developed by the great German chemist Fritz Haber in synthesizing ammonia, Dr. Leroy Spence's catalysis laboratory found ways of inexpensively synthesizing two chemicals with commercial potential: isooctane and dimethylamine. In 1932 the Bristol plant began manufacturing both. Isooctane enjoyed modest but steady sales for use as a reference standard for the measurement of gasoline combustion qualities, the well-known "octane rating." Monomethylamine manufacture first had been suggested to Rohm and Haas by the Tanners Council laboratory, which had discovered that methylamines in general were effective dehairing agents. Rohm and Haas researchers demonstrated that dimethylamine was the most effective. While tanners provided the initial market, methylamines later gained larger sales as accelerators for rubber man-

36

ufacture and intermediates for the synthesis of a wide variety of compounds.

As the Rohm and Haas research effort expanded, the proliferation of laboratories led to logistical problems. By the early 1930s there were labs not only in Bristol and Bridesburg but even at the downtown office. This confusion persisted until 1935, when the first Rohm and Haas central research facility opened—Bridesburg Building 60.

While developing his own products and searching Europe for others, Haas remembered his old friend and partner, Otto Röhm. Haas had promised Röhm he would attempt to reacquire Röhm's original interest in the American company, which was now owned by Tanners Products.[8] Haas kept this promise in 1923 when he reached agreement with Tanners Products to purchase its three thousand shares of Rohm and Haas stock over three years for $500,000, $200,000 more than had been paid. The purchase was completed in two parts, one-third in 1925, and the remainder the following January. Once the purchase had been completed, putting all of the company stock in his hands, Haas used forty percent of it to establish a beneficial trust for Otto Röhm and his heirs.[9] This meant Röhm would henceforth receive the dividend income from the forty percent, while Haas would retain voting control. That summer Röhm made his one and only trip to America and signed the agreement. Now Haas had the legal status of the company where he felt it should be, with himself firmly in control, but with Röhm reaping a proper share of the profits.

The maintenance of close relations with Röhm assured Haas of access to the new products of Röhm's still-fertile mind as well as to those of other scientists in the laboratory of the Darmstadt company. In the 1920s Röhm had resumed work on acrylic acid, its esters and polymers, the topic of his 1901 doctoral dissertation. By 1926 he was well on the way to the development of an industrially practical synthesis for acrylic acid. In 1928 Röhm introduced the first commercial acrylic product, Luglas, a methyl acrylate interlayer for automotive safety glass, to the German market. By 1931 Haas had set up an acrylic lab of his own, and in that year Rohm and Haas Company introduced Plexigum, its own product for the safety glass market. In future years, acrylics would come to play a bigger role in both firms.

By introducing new products and expanding sales of existing ones, Rohm and Haas prospered in the 1920s. Its sales rose from $1.3 million in 1921 to $2.5 million in 1925, $3.7 million in

By 1927 Rohm and Haas needed larger headquarters. It moved into this building at 222 West Washington Square in April of that year. The Washington Square office served as the company's headquarters until 1965.

1927, and $4.2 million in 1929. Net profits soared sixfold to over $600,000. The relative importance of the various lines changed; in 1929 Oropon accounted for but one-seventh of total sales, only slightly more than did syntans. Hydrosulfites became the dominant line, the source of forty percent of Rohm and Haas revenues. The remaining sales came from a variety of products including Orthochrom leather finishes and cuprous oxide, the latter manufactured at the Lennig plant at the request of the U.S. Navy for use in antifouling paints for ship bottoms.[10] As the company grew, many new buildings were constructed at its locations in Bridesburg and Bristol. In April 1927 its corporate offices moved from Front Street to a newly purchased five-story building at 222 West Washington Square.

Only export sales failed to grow during the twenties. Rohm and Haas had picked up business in South America during the war because of the inability of German chemical firms to supply their customers. After the war the German companies resumed their exports, and Rohm and Haas's market collapsed. By the end of 1921, export sales had declined to between $4,000 and $5,000 a month, only a tenth of the wartime level. The following year Haas closed his branches in Buenos Aires and Santiago. Through succeeding years, Rohm and Haas exports consisted solely of small shipments of Oropon to South American chemical brokers.

To a large extent the company's steady domestic growth reflected the general expansiveness of the American industrial economy of the decade. Between 1922 and 1929, productivity, wages, consumer purchases, profits, and dividends, all increased by at least one-fourth. But the prosperity was uneven. Farm income and prices remained depressed, as did soft coal mining and, after 1926, construction.

In 1929 the wall of prosperity came tumbling down. Within a few weeks in October, the stock market lost forty percent of its aggregate value and the country plunged into what would become known as the Great Depression. The depression continued to deepen over the next several years. Between July 1929 and July 1932, the dollar volume of the nation's business declined by fifty-six percent. Unemployment rose to an estimated twenty-five percent, and the country seemed at a loss for solutions.

Although Rohm and Haas certainly could not escape the ravages of hard times, it fared better than most companies. Between 1929 and 1932, sales fell by a quarter and profits by a half, but the average corporation experienced twice as great a decline and many firms went bankrupt. Several factors help explain Rohm and Haas's relative good fortune. One was Haas's long-held insistence on high liquidity and relatively low dividend payouts. This gave him the financial cushion he needed to avoid the drastic cutbacks in new investment and maintenance which many companies found necessary. Rather than allow his physical plant to deteriorate during periods of slack demand, Haas kept his workers on the payroll and had them paint, fix up, and improve the property, thus readying the company for whatever business might develop. With this strategy, Haas made it through the entire depression without laying off a single worker, an achievement of which he was singularly proud and one few firms could claim.

The nature of its markets helped Rohm and Haas achieve this enviable record. Its products went chiefly into two areas, textiles and leather. Within these fields the company's main efforts (hydrosulfites, bates, and syntans) were dominant in their specific market niches. Clothing and shoes are necessities of life, so the decline in these industries was less severe than in industries producing heavier, more expensive, or more durable goods, such as buildings and machinery. The firm sought increased sales through the introduction of improved leather and textile chemicals including dimethylamine for dehairing in 1933 and Primal leather binders in 1934. Lufax zirconium-based opacifier for enamels, a product of Rohm and Haas research introduced in 1933, demonstrated Haas's willingness to pursue new markets even in the midst of the national economic emergency.

Unlike many businessmen, Haas never lost faith in his ability to turn a profit from investment in innovation. If he had a good product and marketed it properly, it would sell. As he noted in his confidential annual report for 1935, "The only satisfactory way to conduct a chemical business is to try to lead the way in research and development." He continued the expansion of his laboratories. Not only did he open the Bridesburg Building 60 research center in 1935, he expanded it twice within the next six years. Although American industry as a whole did not return to 1929 levels of output until 1939, Rohm and Haas surpassed its 1929 sales in 1935. The Washington Square office building, which in 1927 had seemed so spacious that laboratories were installed on the top floor, was now too small. The company bought adjoining properties on two sides to gain additional office space.

Haas no doubt took great pleasure in the state of his company in the mid-1930s. Through good times, and especially through bad, it had consistently outperformed the national economy, growing at a rate that necessitated a steady expansion of facilities. Among these was the strong, productive research division, which already had yielded modest commercial results and promised to do more in the future. Drawing upon both this research effort and products acquired in Germany, Haas promoted diversification of his business. Yet he stayed within the general confines of the field he knew best, the manufacture of standardized, technologically advanced chemicals for industry. The bulk of the company's sales remained in hydrosulfites, bates, and syntans, but Haas could confidently predict that dependence on these lines would lessen over the next decade. What Haas could not have predicted was that the company would be turned upside down and set off in a completely different direction by products emanating from Dr. Röhm's work in acrylic chemistry.

NOTES

1. *Haas to Wallin, April 6, 1921, and April 30, 1921, Rohm and Haas Company archives.*

2. *Thomas Blackadder and Ian Somerville, "The Development of Syntans in the United States,"* Journal of the American Leather Chemists Association 48 (1953): 673–81.

3. *S. Kohn, J. Breedis, and E. Crede, "Comparative Observations of the Tanning Properties of Vegetable Tanning Materials, Synthetic Tans, and Mixtures of Vegetable Tanning Materials with Synthetic Tans,"* Journal of the American Leather Chemists Association 17 (1922): 450–60.

4. *Kohn to Haas, December 21, 1922, Rohm and Haas Company archives.*

5. *Harold G. Turley, interview with M. Biddle, May 27, 1981, copy in Rohm and Haas Company archives.*

6. *Advertisement,* Soap 12 (May 1936): 102.

7. *Donald Murphy, "Brief History of the A&SC Department," August 1948, Rohm and Haas Company archives.*

8. *Agreements between Otto Röhm and Otto Haas, September 9, 1920, Rohm and Haas Company archives.*

9. *Wallin to Haas, September 12, 1923, Rohm and Haas Company archives.*

10. *Rohm and Haas Company, "Annual Report for 1929," Rohm and Haas Company archives.*

THE RESINOUS PRODUCTS
AND CHEMICAL COMPANY

A visitor to 222 West Washington Square in the 1930s would have found a third name on the door beside Rohm and Haas Company and Charles Lennig and Company—The Resinous Products and Chemical Company. If the visitor had asked to see the president of this firm, he would have been sent to Otto Haas. Similarly, if he had needed to see the treasurer, the research director, or a vice president, he would have been referred to men whose names he might recognize as Rohm and Haas executives. If he had reason to visit the plant, he would have found himself on his way to Bridesburg. Yet if he had inquired as to the relationship between Resinous and Rohm and Haas, he would have been told that Resinous was an affiliated company and not a subsidiary. What the visitor may or may not have been told was that Otto Haas owned majority control of both firms.

Resinous Products had its beginning in Haas's visit to Germany in 1924. On this trip he met Drs. Kurt Albert and August Amann of the German firm *Chemische Fabrik Dr. Kurt Albert.* They had invented a series of synthetic resins which they marketed under the trade name Albertol for use in the manufacture of varnish. Traditional resins, such as rosin, damar, and kauri, were hardened tree saps, either fresh or fossilized. Albertol resins were formed by the condensation of a phenol derivative and formaldehyde in the presence of small quantities of natural rosin. They combined some of the advantages of synthetics with some of those of the natural product. Albert offered the Albertol resins to Haas, who agreed to investigate the products' American prospects.[1]

Resins are one of four components used in the manufacture of varnish. The others are drying oil, solvent, and drier. The resin and oil are "cooked" together, then thinned with the solvent. Finally, the drier, a catalyst, is added. The resin provides hardness; the oil, flexibility; the solvent, reduced viscosity; and the drier, curing speed. When the varnish is applied as a thin film on a surface, the solvent evaporates. The resin and oil then cure to a finished hard coat. As complex natural substances, the resins and drying oils used in making varnish in the 1920s varied in composition and results were inconsistent. Different methods of "cooking" the resins and oils also affected the finished varnish. Each of the many varnish manufacturers had its own secret formula. Success was more dependent on the skilled eye of the artisan than the measurement of the scientist.

Haas found the Albertols attractive because they presented a challenge that was similar in many ways to the one Oropon had presented fifteen years earlier. He hoped American varnish manufacturers would embrace Albertols for the same reasons that tanners had adopted Oropon. Albertols were standardized, innovative products which could replace variable, natural ones. The Albertols had been successfully introduced in Europe, and Haas believed that a knowledgeable, technically trained sales force, equipped with these superior products, should produce profitable sales in the United States.

The initial reception of American varnish makers to Albertols was decidedly cool. To Haas they seemed resistant to change and unwilling to see his products' advantages. Their attitude should not have been completely unexpected. The tanners had been similarly hesitant in 1907. Moreover, although Albert had provided Haas with its know-how, neither Haas nor anyone else at his company knew much about the American varnish industry. It was dissimilar in many ways from that in Europe. Tung oil, otherwise known as China wood oil, was the predominant drying oil in America, while linseed oil, which had substantially different properties, was the European standard. Albertols had to be adapted to American varnish industry requirements, a task Haas gave to Dr. Siegfried Kohn. Kohn reported this would require extensive work. Simple substitution of Albertols for natural resins would not produce satisfactory China wood oil varnishes. Entirely new formulas and compounding procedures were required.[2]

Convinced he had a good product, Haas was willing to invest in extensive development. By 1926 Kohn had developed Albertol–China wood oil formulas for different types of var-

E. C. B. Kirsopp, vice president and director of the company and one of Otto Haas's closest aides.

nish. He also had demonstrated that these Albertol-based varnishes were at least as durable as natural resin-based products. Haas, in turn, had learned something of the American varnish industry. Further, the industry itself became more amenable to novelty after accepting a new starting material, nitrocellulose, for use in brush lacquers.

Haas now was ready to begin intensive marketing, but he did not just announce that Rohm and Haas had a new product. Rather, on October 11, he and the Albert company set up a new enterprise, the Resinous Products and Chemical Company, to exploit the Albertols. Haas put up $105,000 capital in return for seventy percent of the stock. The Albert company put up its American patent rights, know-how, and trademarks for the remaining shares. Resinous Products began operations on January 1, 1927, taking over what little business Rohm and Haas had developed. It contracted with Lennig to manufacture the resins at Bridesburg, with Rohm and Haas to provide most other services. The reasons for setting up a separate corporation are not clear, especially since Rohm and Haas had done the development work. Albert may have insisted on an equity interest rather than the more customary license and royalty arrangement. Haas, on the other hand, may have preferred to keep the costs and profits of this new venture separate from Rohm and Haas.

Although Haas was president of the new company, he left much of the day-to-day operations to Edgar (or E.C.B.) Kirsopp. Kirsopp, a Scottish lawyer, had joined Rohm and Haas in 1919 and soon became Otto Haas's most trusted aide. When Haas made his trips to Europe, Kirsopp was left in charge of the company. In 1925 Kirsopp became a vice president and director of Rohm and Haas. When Haas assigned space in the Washington Square building in 1927, he placed Kirsopp's office next to his own. The two offices had an interconnecting door, which rarely was closed.

One of the first things Kirsopp did was change the name of the product line from Albertol to Amberol. This name derived from amber, a high quality, but expensive, natural fossil resin. He hoped such an evocative name would appeal to American varnish makers.

Modest sales of one grade of the resin (Albertol 111 L or Amberol B/S 1) had begun under Rohm and Haas aegis in 1926. One small manufacturer, the George Wetherill Company, used the resin in compounding a floor and interior varnish, which it claimed would dry in four hours, permitting the application of a second coat on the same day. The combination of Amberol and China wood oil first marketed in Wetherill's 4 R varnish represented a breakthrough; no previous product could boast such a short drying time. After discovering that 4 R worked as claimed and was otherwise the equal of older formulas, three large varnish producers picked up on Wetherill's lead late in 1927 and introduced their own quick-drying varnishes.

Within the next year virtually every American varnish manufacturer reformulated its floor and interior varnishes with Amberol resins. While fast drying was the characteristic that fueled the adoption of Amberol, the manufacturers also got a product of superior uniformity to natural-resin-based varnishes. Amberol use soon spread to other clear varnishes and then to colored enamels. Amberol conquered the "shelf goods" market, displacing at least ninety percent of all other resins used in the compounding of varnishes and enamels for package sales to builders, master painters, and handymen.

The stunning success of Amberol was quickly reflected on Resinous Products's financial statements. Sales soared from $173,000 in 1927, mostly in the fourth quarter, to $962,000 in 1928 when 3.8 million pounds of the three grades of Amberol were sold. After paying a $75,000 dividend to its two stockholders, the company still had more than $200,000 in profits to reinvest in the business. Sales further increased in 1929 to $1.3 million on 6.3 million pounds, and the company paid another $75,000 dividend. Less than two and one-half years after its creation, Resinous Products had repaid Haas's entire investment. Out of gratitude for this rapid growth, Haas sold Kirsopp nineteen percent of Resinous Products at par. Haas believed his senior associates best could be motivated and kept in the company if given a direct stake in its business. Beginning in 1927, he offered similar (although smaller) opportunities to several other officials of his companies.

The new venture's growth made keeping up with demand difficult. Resinous purchased a part of the Bridesburg plant site from Lennig in the fall of 1927. By using Lennig's staff and buying utilities and support services from the older company, Resinous had a 450,000-pounds-per-month plant (Building R1) in operation by May 1928. The only problem was that the facility was already too small; work began immediately on an expansion that came on line in 1929. This brought the capacity up to 700,000 pounds per month. The company discovered that the massive aluminum kettles used for preparing the Amberols wore out quickly. After some experimentation, they were replaced with nickel kettles early in 1930. The nickel kettles also proved unsatisfactory; they tended to color the product. Larger, heavier aluminum kettles became the reaction vessels in 1932.

Recruiting and training a staff quickly enough to keep up with market demand proved troublesome. The purchase of services from Rohm and Haas helped but, especially in sales and research, Resinous needed people of its own with backgrounds in coatings. Dr. Kohn transferred from Rohm and Haas to Resinous Products and undertook research on the development of new grades and variants of Amberol. The company hired other chemists both for troubleshooting at the plant and research in a lab at the Washington Square building.

One might assume that with shared ownership, offices, and plant, Resinous Products and Rohm and Haas ran effectively as one company, but this was hardly the case. Competition and jealousy between the two firms could become intense. One employee recalled years later you were either "a Rohm and Haas man" or "a Resinous Products man," and when your work took you to the other side you were considered an intruder.[3] Although the Resinous research chemists worked under the general supervision of the Rohm and Haas research director, they were expected not to exchange ideas with Rohm and Haas chemists.[4]

Resinous Products's phenomenal early growth could not be expected to continue indefinitely. With the saturation of the

market, the beginning of the depression, and the emergence of competition, it slackened off. However, the company continued to make a lot of money. Between 1930 and 1934, sales fluctuated between $1.2 million and $1.6 million per year, varying

Amberol production began in 1927 in Bridesburg. Resinous Products offered varnish manufacturers a standardized, innovative line of products that replaced variable, natural resins and was able to duplicate the kind of success Haas had enjoyed with Oropon.

more with the general state of the economy than with anything else. Net profits remained high, averaging $336,000 annually. Decreased selling prices for Amberol were largely balanced by a steep decline in its raw material costs; rosin in particular was selling well below the cost of production.

Depressed rosin prices in turn reduced the value of rosin-producing pine forests. Haas and Kirsopp took advantage of this by using some of Resinous's substantial cash reserve to purchase eleven thousand acres of Florida pine forest late in 1931. The next year they organized a subsidiary, Southern Resin and Chemical Company, to own and manage the acquisition. Additional acreage was added during the decade. Southern Resin became a captive source for some, but not all, of the rosin needed for Amberol manufacture, and lessened Resinous's dependence on the vagaries of the natural product market.

Several companies marketed competitive rosin-modified phenol-formaldehyde resins in the early 1930s. The best of these products matched Amberol in quality, forcing Resinous Products to cut prices. Resinous also introduced several new grades designed, at least in part, to preempt competition. Amberol 226, which Resinous introduced in 1930, was one such grade. It was also the first Amberol designed in the Resinous laboratory. Kohn developed it specifically for use with China wood oil. Through research and solid technical sales work, Amberols dominated their market throughout the decade.

From the first, Kirsopp recognized that Resinous Products's narrow base was its greatest weakness. There was no assurance that a new resin might not come along that would displace Amberol, as Amberol had displaced the natural resins. But the search for new products that would diversify the business was inhibited by Resinous Products's relationship with Rohm and Haas. Haas saw no logic in his two companies competing with each other. He restricted Kirsopp's efforts to areas where Rohm and Haas was not involved. In the short run, this meant that Kirsopp was limited to developing other products for the general coatings industry (of which varnish manufacture was just a part).

In 1930 Resinous Products introduced its first such effort out of its laboratory, Paraplex plasticizer for nitrocellulose lacquers. Nitrocellulose lacquers required plasticizers because they were too brittle to be used without modification. Paraplex cost more than other plasticizers, but it produced superior flexibility and was a success. Resinous sold 250,000 pounds of Paraplex with a value of $95,000 in 1931. These figures more than doubled in three years, and the plasticizers became second in importance to varnish resins.

Resinous Products introduced a number of other new products for coating manufacture, only one of which achieved substantial sales by mid-decade. This was Amberol 801, which, despite its name, was a rosin-modified maleic acid resin used primarily in nitrocellulose lacquers. It was introduced in 1931. Several less immediately successful products achieved notable sales later in the decade. Duraplex alkyd resins sold steadily once the original compositions developed in-house were replaced by superior formulas purchased from *I. G. Farben*, the giant German chemical company created by the 1926 merger of the six largest German chemical manufacturers.[5] Duraplex resins found their chief uses in industrial coatings, an area never successfully penetrated by Amberols. The I.G. also granted Resinous a license to produce its Membranit alkyd resin emulsions for paint. Resinous sold these under the name Aquaplex, but only with limited success. Among the other coating industry products produced by Resinous in the 1930s were Oilsolate driers for varnish and Uformite urea-formaldehyde resins. Thus, Haas found his two routes to new products, internal research and acquisition of German technology, as profitable in coatings as he had in leather and textile chemistry.

The Uformite resins also proved valuable to the textile industry in making fabrics crush-resistant. Resinous sold Uformite for textiles to Rohm and Haas, which in turn resold the resin to the mills. Such close working relations between the two companies in making and marketing a product might have seemed an ideal partnership, but it proved to be a source of tension, particularly over the intercompany price. Each company's employees believed the price to be tilted to the benefit of the other.[6]

Kirsopp and Haas found one market where the Resinous expertise could be used to good advantage without stepping on Rohm and Haas toes, and that was plywood adhesives. In 1934 Kirsopp acquired the American rights to Tego gluefilm for plywood from its originators, *Theodore Goldschmidt A. G.* of Essen, Germany. Like many other products introduced by Haas, Tego offered a synthetic, standardized alternative to variable natural products. Plywood is made by gluing together several thin wood veneers under pressure. Previous plywood adhesives, such natural substances as blood albumin, casein, and soybean meal, were applied to the veneer as powders and liquids. Tego was very thin paper impregnated with phenol-formaldehyde resins and was manufactured in long rolls. Sheets could be cut to size by the customer, placed between veneers, and then cured in a hot press. Like many advanced products, it cost more, but produced a better product. Not only did it give more uniform results, it produced a plywood with previously unequaled levels of water and weather resistance. Tego did have one disadvantage: its use required a hot press, which many plywood manufacturers did not own. Nevertheless, Tego achieved sales of $200,000 in 1936. Within two years, however, sales had declined by half.

Resinous itself was responsible for much of this decline through its 1938 introduction of Uformite 430 liquid urea-formaldehyde plywood adhesive. Uformite 430 gained a broader acceptance than Tego because it gave plywood manufacturers the advantages of standardization at a much lower cost and could be employed using the same procedures as the traditional natural plywood adhesives. Several other companies placed similar products on the market in the late 1930s. The competition depressed prices, but the two-year lead Tego gave Resinous Products with the plywood companies helped the firm remain a leader in this field. Tego retained a limited but important market in the highest grade plywoods, especially those for aircraft and marine applications.

The success of the Resinous Products and Chemical Company in the coating and plywood adhesive fields pleased Haas as much as the continued growth of Rohm and Haas itself. It demonstrated that the strategies and techniques that worked for Haas in leather and textiles could lead to success elsewhere. Resinous, like Rohm and Haas, was committed to technologically advanced, standardized products for industry. It sold these through technically trained salesmen who could work to solve the customers' problems. From a single product, Resinous had expanded to encompass a wide range of successful chemicals for the coating and plywood industries, and it had done so in spite of the Great Depression. Haas could feel confident of the future prosperity of both of his enterprises.

NOTES

Additional material for this chapter is derived from the unpublished annual reports of the Resinous Products and Chemical Company in the Rohm and Haas Company archives.

1. Ernst Schwenk, 125 Jahre Albert Chemie im Biebrich am Rhein *(Weisbaden: Hoechst Albert Weisbaden, 1983), 58–64.*

2. Siegfried Kohn, "Progress Report for 1924," "Progress Report for 1925," Rohm and Haas Company archives.

3. George Schnabel, interview with the author, June 14, 1983, copy in Rohm and Haas Company archives.

4. Ralph Connor, interview with the author, September 29, 1983, copy in Rohm and Haas Company archives.

5. Joseph Borkin, The Crime and Punishment of I. G. Farben (New York: The Free Press, 1978).

6. Frederick interview.

PART TWO ◂

REDIRECTION AND EXPANSION 1934 ⋈ 1959

The cover for Otto Röhm's doctoral thesis, "On the Polymerization Products of Acrylic Acid"

Ueber

Polymerisationsprodukte der Akrylsäure.

Inaugural-Dissertation

zur

Erlangung der Doktorwürde

einer

hohen naturwissenschaftlichen Fakultät

der

Eberhard-Karls-Universität zu Tübingen

vorgelegt

von

Otto Röhm

aus Ohringen.

Tübingen.

Verlag von Franz Pietzcker.

1901.

CHAPTER 5

THE RISE OF ACRYLIC CHEMISTRY

The Rohm and Haas companies of Darmstadt and Philadelphia had been founded on Otto Röhm's abilities and discoveries in enzymology, but his primary training had been in the more classic areas of organic reaction and synthesis. His doctoral dissertation, "On the Polymerization Products of Acrylic Acid," had been in this field. Röhm was particularly interested in the acid's esters (the products of its reactions with alcohols) and in the self-reaction products of these esters, what today would be called acrylate polymers. These polymers had a combination of clarity, toughness, and flexibility that suggested to Röhm a cross between glass and rubber. From his school days on, he believed acrylic acid derivatives had commercial potential, but he also knew such commercialization would require extensive research.

The demands of building and running the business in Darmstadt limited the effort Röhm could devote to acrylics. Between 1912 and the outbreak of war in 1914, however, he found time to obtain patents on acrylic ester polymers as rubber substitutes and replacements for drying oils in varnishes. Even then, Röhm could not market anything based on these patents because he lacked the knowledge needed for large-scale production. The synthesis Röhm used for production of monomers required expensive starting materials, took many steps to complete, and returned low yields.

World War I and the disruptions caused by the German defeat slowed Röhm's acrylic research, but around 1920 he resumed this work with a new assistant, Dr. Walter Bauer. Together they developed a new synthesis for acrylic esters using ethylene cyanohydrin as the starting material. This procedure was a step in the right direction, as ethylene cyanohydrin could be made from commercially available ethylene chlorohydrin. Chlorohydrin had been developed as a starting material for chemical warfare agents. It took Röhm and Bauer until 1926 to turn their synthesis into a commercial process.

At the same time, Röhm and his staff sought commercial applications for the polymers of the two simplest esters, methyl and ethyl acrylate. In 1927 they developed one which involved the use of polymethyl acrylate in automotive safety glass. Automotive safety glass is a sandwich of clear plastic between two panes of glass. Polymethyl acrylate formed a more satisfactory interlayer than the nitrocellulose then in use. Acrylate did not yellow with age, was less flammable, and produced a glass with greater splinter resistance. Darmstadt introduced this first acrylic product in 1928 as Luglas. Luglas was formed by spreading a layer of a polymethyl acrylate solution on a glass plate, allowing the solvent to dry, and then covering it with a second glass plate that had been coated with a plasticizer. Luglas enjoyed a good initial reception, but sales slowed as the depression forced a steep decline in automobile manufacture.

Röhm kept his partner Otto Haas posted on the progress of Darmstadt's acrylic research. As early as 1929, Haas's chemists were themselves investigating acrylic syntheses,[1] and in 1930 Haas established his own acrylic research laboratory under Dr. Harry Neher. Neher regularly exchanged data and ideas with Darmstadt. This led to a 1931 agreement between Haas and Röhm in which Rohm and Haas Company agreed to subsidize Röhm and Haas A.G.'s acrylic research in exchange for the North American rights to its results.[2] In the same year, Haas, working in conjunction with the American Window Glass Company, introduced his own version of automotive safety glass. He called his interlayer Plexigum.

Luglas and Plexigum enjoyed a brief, profitable heyday before losing their markets to technically superior, less expensive polyvinyl acetal and polyvinyl butyral films in the late 1930s. The acrylic safety layers had at least two shortcomings. They performed poorly at low temperatures and their soft, rubbery consistency made them difficult to cut into various shapes and sizes. For this reason, both Luglas and Plexigum safety glass had to be manufactured in the exact sizes needed.

Röhm's laboratory continued its exploratory research, and in 1931 Bauer presented him with a sample of polymethyl methacrylate, the polymer of the methyl ester of methacrylic acid. Polymethyl methacrylate is a transparent solid which softens only above 110° C. Unlike the softer acrylates, it can be easily worked with standard tools such as saws and drills. It struck Röhm as a far more useful material than the acrylates he had been working with.

Between the general economic depression and the expenses of his ambitious research program, Röhm was strapped for both cash and resources. He negotiated a licensing agreement whereby *I. G. Farben* took over acrylate production and development. This freed Röhm's assets for methacrylate research and promised him future royalties as the I.G. used its immense resources to explore and to promote additional markets for acrylate polymers. Röhm retained only the production of safety glass for his company.

The Darmstadt and Philadelphia scientists searched continually for means of overcoming the technical shortcomings of their acrylate safety glass, even while much of their laboratory effort shifted to methacrylates. Röhm and Bauer knew different polymerization conditions could produce resins with different properties, so they tried several alternatives, including insite polymerization. They made a cell out of two glass planes by lining them up in parallel, with a spacer of stiffened coated paper around the edges in between the glass planes. They poured monomer into the cell and allowed polymerization to take place. Bauer tried methyl methacrylate as the monomer in this glass cell system in 1932 and a strange thing happened. When the sandwich was cooled, the glass did not adhere—it cleaved cleanly revealing a solid, transparent, colorless polymer sheet. Röhm quickly realized this was a major breakthrough, a discovery with far greater potential than any auto safety glass. Röhm and Bauer had invented an entirely new material; to use Röhm's phrase, they had produced the first "organic glass."

A laboratory curiosity is one thing; a commercial product is something else. Röhm and his assistants spent the next three years developing this cast acrylic sheet from an experimental to a production process and investigating the properties and potentials of the sheet. Röhm recognized the impracticality of glass for the cell walls but an extensive search failed to develop an alternative system.[3] Another problem was caused by heat from the polymerization and its effect on the remaining monomer. Darmstadt solved this by placing the cells in ovens which controlled the rate of polymerization. The researchers also developed new catalysts to initiate the polymerization.[4]

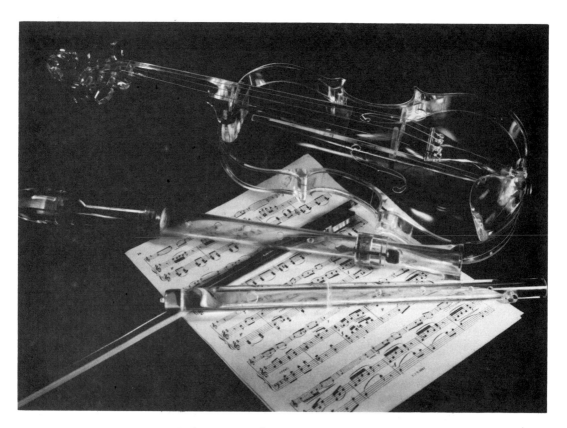

Röhm's researchers experimented with many different uses for their new acrylic material, including musical instruments like this violin and flute.

The prospects for success so impressed Haas that he set up his own methacrylate laboratory in Bristol in 1934. Reports began flying fast and furious in both directions across the Atlantic.

The manufacture of cast methacrylic sheets required a process for large-scale methyl methacrylate monomer manufacture. Röhm's lab designed a satisfactory variant on the acrylate monomer technique, which went into production in Darmstadt in 1933. Two years later Röhm learned that Imperial Chemical Industries (ICI), the dominant British chemical manufacturer, had been working on the same problem and had developed a cheaper, shorter process. Röhm was able to obtain a license to use the process in Germany by offering ICI a license to make cast sheet in Great Britain. Haas was unable to obtain a similar license for the United States until the mid-1940s and thus had to use the original technique.

In addition to learning how to make sheet, Röhm and his associates had to discover applications for which it could be sold. Röhm took a personal interest in this investigation. He wore what was undoubtedly the first pair of plastic spectacle lenses. He replaced the side windows (successfully) and front windshield (unsuccessfully) of his automobile with acrylic sheet. His researchers produced an acrylic violin, which did not have a good tone, and an acrylic flute, which did. These novelties demonstrated the techniques Darmstadt was developing for forming the sheets into three-dimensional shapes. Polymethyl methacrylate is thermoplastic, which means it softens when heated and hardens when cooled. Sheets could be shaped easily by heating them to about 110° C, after which they could be stretched over a form and hardened by cooling. By the beginning of 1936, Röhm was ready to announce commercial availability of cast polymethyl methacrylate sheet, to which he gave the trade name Plexiglas.

Although Rohm and Haas of Philadelphia had been working on acrylic sheet since 1934, it still knew less than Darmstadt. Therefore, as Darmstadt commenced production in the winter of 1935–36, Haas sent Dr. Donald Frederick, a young chemist from his plastics lab, to Germany to learn the art of Plexiglas manufacture. Even though the Nazi government had begun placing restrictions on the transmission of technical reports

The German aircraft industry was testing Plexiglas for use in military airplanes in the mid-1930s. Plexiglas was used for cockpit enclosures in planes like this Heinkel-Flugzeug He 111 K.

abroad, it was still possible, apparently, to transfer information in person. In any case, know-how is often best taught directly.

Frederick spent two months in Germany, observing, working in the plant, and holding discussions with both research and production men. He learned everything Darmstadt knew about the complex process of manufacturing Plexiglas sheet. What he did not see, he realized later, was much work on the forming of flat sheet into three-dimensional shapes. Most likely the Nazis did not want an alien to observe the production of parts intended for military aircraft in the rearmament program they had begun in violation of the treaty of Versailles.[5] Frederick knew the Germans were experimenting with Plexiglas windows in their airplanes, and its usefulness for aircraft must have been obvious to him from the characteristics of the material alone. It was lighter in weight and more transparent than glass, formable into three-dimensional shapes with mild heating, able to withstand pressure differentials, weather-resistant, and shatter-resistant (a bullet would produce a clean hole).

Frederick returned to Philadelphia at the end of February 1936 and, as planned, transferred from research to direct the sales of the new product. Rohm and Haas opened its first small Plexiglas production line at the Bristol plant in April, and the following October issued its first Plexiglas price list. The company offered sheet of thicknesses from 0.060 to 0.240 inches, and dimensions of up to 36 by 48 inches. Total sales for 1936 were under 4,000 square feet, or $13,000.

Frederick also spent considerable time in the field looking for markets for the new material. Among the outlets he pursued were companies that made spectacles, instrument covers, dentures, display cases, and lighting fixtures. Some of these uses gained public attention in 1939 when they were exhibited at the New York World's Fair. In addition to Rohm and Haas's own small display, Plexiglas figured prominently at the General Motors building, which featured an illuminated Plexiglas sign and a Pontiac with a Plexiglas body that showed off the car's inner workings. In seeking such markets for Plexiglas, Frederick followed Haas's basic business principle: making and selling materials to be used in the manufacture of finished goods.

Plexiglas found a major market almost immediately in military aircraft. In 1938 Rohm and Haas sold $431,577 worth of sheet and more than $400,000 of this went into airplanes. Frederick had paved the way for these sales in 1936 by convincing the Army Air Corps Command at Wright Field in Dayton, Ohio, to issue specifications authorizing the use of methyl

During the war, Rohm and Haas employees at the Bristol plant not only produced Plexiglas, but also fabricated the material into cockpit canopies, windshields, and so on, needed to meet military demands.

methacrylate, and no other plastic sheet, in the construction of military aircraft. Before taking this action the Army Air Corps insisted that Rohm and Haas physicists investigate the physical properties of Plexiglas. The company opened a lab for this purpose and generated numerical test data that the Army could use to determine the suitability of Plexiglas for military use. The process took several months. It was then repeated for the Navy Air Corps.

Army and Navy approval went far toward assuring acrylic sheet's acceptance by the manufacturers of military aircraft. The aircraft industry's technological sophistication was advancing rapidly, and a manufacturer who wished to obtain contracts had to keep up.[6] Over the next two years, Plexiglas replaced glass in the windows of military planes. While Rohm and Haas's 1936 Plexiglas brochure featured the company logo on the cover, and the first printing of the April 1937 brochure displayed a selection of Plexiglas shapes, the second 1937 printing spotlighted a Seversky P35 fighter plane with a Plexiglas canopy, the first of many such airplanes to be featured. While Plexiglas would find other uses during its first decade, its future was entwined with that of the aircraft industry.

The earliest aircraft incorporating Plexiglas used it largely as glass, in small, flat panes. As manufacturers became more familiar with the material, they altered aircraft design to exploit the unique properties of Plexiglas. It encouraged the growing trend toward streamlining; open cockpits disappeared, and planar windows gave way to three-dimensional transparent shapes. Larger planes sprouted transparent 360° gunner's turrets and bombardier's enclosures. Plexiglas gained this market quickly but it was not, at first, a large one. In 1937 American

Plexiglas sheet formed the bombardier and gunner enclosures on the nose, the gunner turret on top, the side windows, and tail assembly of this giant Douglas B-19.

military aircraft manufacturers sold 949 planes; two years later, they produced 2,141.[7]

This relatively small market gave Rohm and Haas needed time to improve both process and product. The company learned much about the performance of Plexiglas under actual conditions of use. Its laboratory found, for example, the fabrication of parts could be improved by overheating the sheet and letting it partially cool before stretching it over a form. The lab began experimenting with a vacuum molding technique which produced shapes of more uniform thickness but caused, initially, a decrease in optical quality. Aircraft manufacturers discovered that a Plexiglas part affixed to a plane via a rubber-lined channel resisted cracking better than one attached with screws or rivets. Rohm and Haas's technical sales service force disseminated the steadily increasing knowledge of Plexiglas characteristics to the company's many acrylic customers. The Philadelphia and Bristol labs did not work alone in these improvements. Reports were exchanged regularly with Darmstadt. Haas provided further details as the result of his annual trips to Germany.

In the late 1930s, Haas sensed the coming of war and, believing he had an essential military product, he expanded his firm's sheet capacity from 5,000 sheets per month in January 1937 to 20,000 sheets per month in January 1939 and 70,000 in January 1940. This commitment, amounting to an investment of over $1 million in facilities alone by 1940, was expensive for a company with total sales under $10 million. Haas recognized that war or not, his company was changing in both size and direction. For the first time he could count among his customers some of the titans of American industry, companies like General Motors. He was playing a larger game with greater potential for both risks and rewards.

As World War II began, American aircraft production increased dramatically. The 2,141 military planes produced in 1939 became 6,019 in 1940, and, with the passage of the Lend-Lease Act, 19,453 in 1941. Rohm and Haas continued to keep pace—and then some. Despite price declines of nearly fifty percent, the company sold $4.5 million (or 2.2 million square feet) of sheet to the aircraft industry alone in 1941. These sales made plastics the largest division of the company. Although the government repeatedly assured Haas that his capacity plus that of the smaller plant of Du Pont, which had begun manufacturing cast sheet in 1939, would suffice for future demands, he continued to expand his plant. By the end of 1941, manufacturing capacity reached an unbelievable 386,000 square feet per month, representing a total investment of close to $5 million.

Haas's foresight enabled Rohm and Haas to keep abreast of the demand, but he was not satisfied. Convinced that neither the aircraft industry nor subcontractors would be able to keep up with the demand for finished parts, he established fabricating plants at Bristol, Pennsylvania, and in July 1941 at Southgate, California. The new plants marked a radical departure for the company because they were its first entry into finished products. Haas opened them solely to meet the war emergency and closed them as quickly as possible in 1945 when demand slowed.

By September 1942 he had expanded the sheet capacity of the Bristol plant to over a half-million square feet per month. Aircraft construction increased apace. Nearly 48,000 planes rolled off the lines in 1942, with still higher amounts predicted for the following years. With these predictions in mind, Haas convinced the government in 1942 that although an entirely new acrylic sheet plant was needed, the project was beyond his already overtaxed financial resources. The Defense Plant Corporation then financed the conversion of an abandoned furniture factory in Knoxville, Tennessee, to Plexiglas production. The new plant produced its first sheet in March 1943, a scant eighteen weeks after groundbreaking. This came just in time; 1943 saw the United States production of 86,000 warplanes.

Many technical problems resulted from this rapid expansion. Plexiglas grew so big, so fast, no one understood what could go wrong, especially during sheet manufacture. Rohm and Haas had a production schedule that allowed no time for slowdowns or shutdowns; yet, considering the novelty and complexity of Plexiglas manufacture, the demands of this schedule were an open invitation to problems. At one point a majority of the sheets produced were useless due to the appearance of pimples, circular surface depressions surrounded by raised rings. After getting no results from several sophisticated chemical tests, one researcher examined a pimpled sheet under a microscope and discovered a minute metallic particle at the center of each pimple. A search of the production line discovered this metal in the solder of the funnels used to pour the monomer into the cells. The funnels were replaced, and the problem was solved. Among other problems the lab overcame were haziness, smear patterns, and sheets that shattered immediately upon removal from the cell.[8]

In the early part of the war, fabricators made parts by heating the sheet and manually stretching it over a form; shaping a large piece could easily require four men. By 1944 the Rohm and Haas laboratory could demonstrate blow and vacuum forming techniques which produced needed three-dimensional shapes more easily and with less expense.[9] It is often noted that war serves as an accelerator of technological change; the history of Plexiglas is a sterling example of this maxim. Plexiglas led Rohm and Haas through a series of separate learning experiences on a complex process that, beginning with a few raw materials, finished with an aircraft part.

The wartime transformation of Rohm and Haas's business operations and orientation was as dramatic as the technological changes. In 1944 Rohm and Haas sold over $22 million worth of Plexiglas sheet and an additional $6.5 million of other acrylic products. Virtually all of this production went into military uses. Altogether, acrylics accounted for two-thirds of total sales, and total sales had multiplied to many times their prewar level, from $5.5 million in 1938 and $8.2 million in 1939 to $43 million in 1944. Similarly the company's workforce had increased from one thousand to over seven thousand employees. The old standbys—bates, syntans, and hydrosulfites—were still there and selling in customary or somewhat increased amounts, but they were dwarfed by Plexiglas. The magnitude of the market available to acrylics was just many times greater. The small manufacturer of leather and textile specialties had become the fountainhead of acrylic plastics, a new and important basic material.

These changes caused problems throughout the organization. Up to this point, Haas had been able to keep very close tabs on everything that went on in his companies. He made all major decisions himself and, with the exception of Kirsopp at Resinous Products, he rarely delegated authority to other executives. But Rohm and Haas had grown too big for such one-man rule. Moreover, Haas reached his seventieth birthday in 1942 and, although still vigorous and showing no signs of slowing down, he knew it was wise to give younger men broader responsibilities. An accident in February 1943 which kept him away from the office for several months reinforced this belief.

In 1939 he had begun planning a new organizational structure, one which would be divided into several functional departments. Over the next four years he built his management team. From within came Donald Frederick in sales, Louis Klein in Resinous Products sales, and Dr. Lloyd Covert (formerly a research chemist) in production and construction. He also re-

Lloyd Covert was named to head up production and construction as part of Otto Haas's effort to introduce a more detailed organizational structure.

Louis Klein was named to head Resinous Products sales.

Donald Frederick was named vice president for sales in 1943.

cruited a financial expert, Duncan Merriwether, from New York's Irving Trust Company, and a purchasing agent, Dr. Peter Clarke, from the New York University School of Business.

In 1943 Haas promoted Frederick to the new position of vice president for sales. While previously there had been a separate sales organization reporting to Haas for each trade Rohm and Haas served, now all sales operations would report to Frederick. Similarly, Klein became vice president of Resinous Products, in charge of that affiliate's sales. Merriwether became treasurer and chief financial officer. (Previously, Van Wallin, a holdover from the Tanners Products era, had held the title of treasurer but had never actually performed as such.) Covert became assistant to the president and the next year a vice president with jurisdiction over engineering, construction, and production.

Ralph Connor was hired in 1944 to oversee the research organization.

Duncan Merriwether, hired from New York's Irving Trust Company, was brought into Rohm and Haas because of his financial expertise.

The new organization was completed in 1945 when Haas
hired Dr. Ralph Connor to supervise the increasingly crucial
research organization. During the war Connor, who had
formerly taught chemistry at the University of Pennsylvania,
had directed a federal division of explosives research. He became Rohm and Haas's vice president for research in 1948.
With these men in place, Kirsopp gradually reduced his role
within the organization; after 1948 he occupied only the largely
honorary position of vice chairman of the board of directors.

The departmentalized organizational structure Haas set up
in 1943 was then typical of companies of Rohm and Haas's
size.[10] But to a large extent it existed only on paper at Rohm
and Haas. More of the routine and semi-routine decisions were
being made elsewhere, but Haas himself continued to provide
overall direction and planning and, to the extent that time
allowed, continued to watch even the most routine details. The
new executives referred to Haas decisions that in most other
companies they would have been expected to make themselves.

As the number of executives at the Washington Square office grew, the need for more office space became acute. In 1945
Rohm and Haas purchased a larger office building some ten
blocks west at 1700 Walnut Street with the announced intention of moving its headquarters there. This decision was re-
versed within the year, perhaps because Haas did not want to
leave his park-view office. Several departments, including accounting and purchasing, moved into the new building, however, and for the next twenty years successive generations of
mail clerks and couriers trod the blocks between the two offices.

The war transformed Rohm and Haas in at least one other
way. For the first time the company was completely dependent
on its own research facilities for development of new products
and improvements to existing ones. Even while building his
own research division, Haas had maintained close ties and
business relationships with Darmstadt and the rest of the German chemical industry. He had continued his annual business
trips to Germany until the actual outbreak of war, visiting as
late as August 1939, the month before Germany invaded Po-
land.

But business was not the main motivation for Haas's 1939
trip. Rather, he went to Darmstadt at the request of friends
who hoped he would be able to shake Röhm out of a deepening
depression which had begun with the death of his wife in 1936.
While he was in Germany, Haas held extensive discussions

with Darmstadt scientists and with scientists and top officials
of the I.G. Whatever success this trip had from a business
standpoint was clouded by the continuing despondency of Otto
Röhm. One month after Haas returned to Philadelphia, his old
friend and partner, with whom he had shared so much, died in
Darmstadt.

The research division Haas had been cultivating carefully
since the 1920s began repaying his investment. It proved equal
to the demands placed on it. While the plastics production lab
was meeting the immediate problem of keeping first-quality
Plexiglas pouring off the lines, the acrylics research team
produced several improved copolymer formulas for Plexiglas.
Darmstadt had used ethyl acrylate as a plasticizer to increase
sheet formability. The research team developed a formula that
replaced this with lesser proportions of a related monomer,
butyl lactate. Butyl lactate was adopted by the production department quite literally overnight when an explosion at Bristol
destroyed the company's main facility for the production of
ethyl acrylate. Still later, the lab devised a practical method for
the manufacture of pure methyl methacrylate sheet. Not every
investigation was quite so successful; extensive work went into
attempted improvements in the abrasion-resistance of Plexiglas, but without much success.[11] To this day, the ease with
which Plexiglas can be scratched is one of the material's limitations.

Rohm and Haas had several scientists working on projects
of a more pioneering nature. These men also developed products of importance to the war effort. The foremost of these
explorers into the chemical unknown was Dr. Herman Bruson.
A native of Middletown, Ohio, he had joined Rohm and Haas
in 1928 at the age of twenty-seven after earning degrees at the
Massachusetts Institute of Technology and the Zurich Poly-
technic Institute. In his first years with the company, he invented Resinous Products's Oilsolate driers and Paraplex
plasticizers. Bruson established a pattern with these discoveries which he would follow throughout his career. Once he
found something interesting, he would apply for a patent and
then lose interest. Someone else would have to take the discovery from the lab bench to the semiworks and the factory. Bruson would be off trying a new modification of the molecule or
working on an entirely new project. Even in all his pioneering
work, however, Bruson exemplified the difference between industrial and academic chemical research. His goal was to make
discoveries that might lead to useful products, not just to in-

crease chemical knowledge. He was not constrained by niceties such as the division between Resinous Products and Rohm and Haas; he went where his intellect took him.

Not surprisingly, considering he was working at Rohm and Haas, this took him to methacrylic acid in 1934. Röhm, and everyone else working in the field, was looking at the esters formed by the reaction of the acid with simple alcohols like methanol and ethanol. Bruson examined the esters of the higher alcohols, straight chains of eight to sixteen carbon atoms with a single hydroxyl (or alcohol) group on the end. The polymers he formed from these esters were waxy solids. Bruson discovered they had one very peculiar property: when dissolved in mineral oil, they flattened the oil's viscosity index curve without appreciably thickening the oil. Oil containing these esters retained useful thicknesses and flow rates over a greatly extended range of temperatures. Bruson applied for a patent on these esters as oil additives in 1934 and then moved on to other work. Four years later his research was picked up by Dr. Harry Neher, head of the Rohm and Haas acrylics laboratory, who developed several practical formulas between 1938 and 1940. But the phenomenon still remained a laboratory curiosity. The high cost of producing these oil additives made commercial development unlikely.[12]

Then came World War II and suddenly Bruson's discovery took on vast importance. Among the many committees the government set up in the early days of the war was one whose job it was to go through patent files looking for inventions that might aid the war effort. The committee fixed upon Bruson's oil additive patent and sent Rohm and Haas a request for several pounds of the material for testing. Besides the committee's interest, the Army Air Corps headquarters (with which Rohm and Haas was intimately familiar as a result of Plexiglas) assured Haas that the lack of a hydraulic fluid that would remain functional at an ambient temperature range of at least $-50°$ F to $150°$ F ($-45°$ C to $65°$ C) was one of the most critical constraints on military aircraft design. Rohm and Haas, with the assistance of the National Defense Research Committee, began a crash program for commercialization of Bruson's esters. Rohm and Haas shipped its first batches of the resulting product, Acryloid HF, to several manufacturers of military grade hydraulic fluids in 1942. Two formulations designed for lubricating oils followed shortly thereafter. The Acryloid oil additives quickly gained acceptance throughout the military and became Rohm and Haas's second most important contribution

Herman Bruson, an inquisitive researcher, made the discoveries that led to Acryloid oil additives.

to the war effort. In some ways the oil additives were a more striking one, for they were exclusively the product of Rohm and Haas Company research. Military sales of Acryloids peaked in 1944 at $3.3 million, representing close to three million pounds of the additives.

In later years Bruson delighted in retelling the story of the development of the oil Acryloids. He would always add one fillip: he claimed that, through the Acryloids, he was responsible for the Russian victory at the battle of Stalingrad. The Russian tanks and artillery had Acryloid-modified hydraulic fluids and functioned in the depths of the Russian winter; the Germans had only ordinary fluids, and their equipment froze.

Bruson made at least one other discovery, in a completely different area, that contributed to the war effort. This was Hyamine 1622, Rohm and Haas's first bactericide, which Bruson discovered in 1939. It had widespread use for the maintenance of antiseptic conditions in mobile military hospitals.

Bruson was not the only Rohm and Haas chemist whose research aided the war effort. Dr. Harold Turley developed Orotan TV, the first syntan that could produce superior quality leather when used as the lone tanning agent. It served as a substitute for scarce imported vegetable tans. Dr. Leroy Spence contributed a catalytic method for the preparation of acrylonitrile, a chemical that was commercially unknown before 1940 when Rohm and Haas began its manufacture. It became vital to the war effort as one of the two starting materials for the production of Buna N synthetic rubber, which was used for the lining of gas and other fuel tanks. Synthetic rubber was important because the Japanese had cut off American access to the natural rubber plantations of the Far East. Rohm and Haas made millions of pounds of acrylonitrile for the war effort but discontinued production at the war's end. Other companies devised superior syntheses, and Rohm and Haas research failed to uncover any sufficiently attractive peacetime uses for the chemical.

Wartime conditions gave renewed prominence to some older products. Lethane sales increased fivefold to over $3.5 million, thanks in part to dwindling supplies of competitive imported products. These included the natural insecticides pyrethrum and rotenone. Lethane became the sole active ingredient in leading household fly sprays and even gained use on vegetable crops such as cabbage. Other products suffered as allocation systems tied the availability of many needed raw materials to their end uses. Applications for civilian market sales drew low priorities from the authorities.

This allocation system brought similar problems and opportunities to Resinous Products. Sales of some products, most notably Tego gluefilm, grew dramatically. Tego was in heavy demand for aircraft plywood, especially during the early part of the war. Resinous constructed a second Tego line at the Bridesburg plant in 1942, which it abandoned two years later when demand for aircraft plywood declined. Other products remained in tight supply. Not only was allocation of raw materials tied to end use of the product but, from 1942, the Resinous Plant at Bridesburg was operating at close to full capacity, shifting from one product to another as raw material availability and military demands required. Other than Tego, its products did not have sufficient priority with the government to win an allocation of construction material or manufacturing equipment. In the immediate prewar period, 1938–40, Resinous Products's sales had doubled, reaching $5.1 million, and during

the war they doubled again, but this growth came from expanded sales of existing products rather than from innovations.

Tego played a part in construction of one of the most noto-
rious aircraft in military history, Howard Hughes's experimental troop carrier H-4, better known as the Spruce Goose. This
was the largest wooden airplane ever built, but it was never
used. After one brief test flight in 1944, it languished for
decades in a hangar. Naturally, the plywood from which it was
made had been bonded with Resinous Products's gluefilm.

Acrylic chemistry and World War II dramatically altered the
lines of business at Rohm and Haas. In the space of a few
frenzied years, the small manufacturer of specialty chemicals
for the leather and textile industries was transformed into a far
larger company making a new class of materials, acrylic polymers, whose full potential was still unknown. Haas had been
propelled by events into a maelstrom he had not foreseen. To
his credit he had piloted his ship to the calm at the other side
of the storm. It emerged a more formidable but less familiar
vessel, and it was far from certain how well it would handle the
weather ahead. It remained to be seen whether Haas's business
strategies and acumen would prove equal to the new challenges; whether the methods that proved so successful in a
small leather, textile, and coating chemical business would
prove equal to acrylic plastics. What could he do in peacetime
with all that capacity for monomer and sheet?

But even as Haas was preparing for the challenges facing
him in postwar America, he was troubled by challenges of a
very different sort. Haas found himself beset by legal problems
which would have to be resolved so that he could get on with
the business at hand.

NOTES

1. *C. E. Andrews, Report 13, February 25, 1929, Rohm and Haas
Company archives.*
2. *Draft testimony of Otto Haas, May 1945, Rohm and Haas
Company archives.*
3. *Trommsdorf,* Otto Röhm, *187–256.*
4. *Donald S. Frederick, report, February 6, 1936, Rohm and Haas
Company archives.*
5. *Frederick interview.*
6. *Irving Holley,* Buying Aircraft: Materiel Procurement for the
Army Air Forces *(Washington: Office of the Chief of Military History,
Department of the Army, 1964), pp. 6–32.*
7. *Holley, pp. 11, 555.*
8. *Stanton Kelton, Jr., interview with the author, June 24, 1983,
copy in Rohm and Haas Company archives. Letter, O. Haas to Ralph*

Hall, War Production Board, June 19, 1942, Rohm and Haas Company archives.

9. Fred Williamson, "Standardized Eye Section for B-29," Modern Plastics 148 (December 1944): 200–202.

10. Alfred Chandler, Jr., Strategy and Structure (Cambridge, Mass.: MIT Press, 1962).

11. Kelton interview.

12. Herman Bruson, various A Reports 1928–1934, Records of the Research Division, Rohm and Haas Company, Spring House, Pennsylvania. Harry Neher, various A Reports 1938–1940, Records of the Research Division, Rohm and Haas Company, Spring House, Pennsylvania.

6

LEGAL COMPLICATIONS

Rohm and Haas Company, Otto Haas, Donald Frederick, E. C. B. Kirsopp, as well as E. I. du Pont de Nemours Company (the largest United States chemical firm) and five of its executives were indicted in Federal District Court in Newark, New Jersey, on August 10, 1942, on criminal charges of conspiring to control the marketing, production, and price of acrylic products in violation of the Sherman Antitrust Act of 1890. ICI of Britain, *I. G. Farben* of Germany, and Röhm and Haas A.G. (Darmstadt) were named as unindicted co-conspirators. A second indictment, returned the same day against the two American companies and their officials along with several other firms, alleged an additional conspiracy to control the price and marketing of acrylic dentures.

The government's central argument in the case, first presented to a congressional committee in April by Special Assistant to the Attorney General Walter Hutchinson, was that a series of separate two-party patent licensing and cross-licensing agreements, each possibly legal at face value, had cumulatively amounted to an illegal conspiracy.[1]

When word of this impending indictment reached Otto Haas, he was incredulous. He did not believe he had done anything wrong. Quite the opposite, he thought he had gained for the United States the ability to produce a vital war material, Plexiglas, which otherwise would have been unavailable. Moreover, as Stanton Kelton, Sr. argued before the congressional committee, the agreements Haas had signed were based on patents, and patents were legally sanctioned, limited monopolies which governments granted to inventors.

Haas had signed the agreements with Darmstadt and the
I.G. to acquire rights to certain patents owned by the German
companies. These agreements contained terms designed to
protect the patent holder: they restricted Haas to marketing
the products in the United States (and in some instances Canada); required Haas to pay royalties; and in the case of the I.G.
agreement, limited the fields in which he could sell acrylics.
Safety glass, adhesives, and glass substitutes were reserved for
Haas, while synthetic rubber, photographic articles, and pharmaceuticals were reserved for the I.G. Haas gave Darmstadt
exclusive European rights to all acrylic patents to come out of
Philadelphia's labs, and the two Rohm and Haas companies
had long agreed, orally in 1920 and formally in 1927, not to
enter each other's markets. The question of antitrust law had
been recognized when the agreements were drafted. Haas's
lawyers advised him that the agreements did not violate the
Sherman Act, that patents had longstanding legal sanction, and
no antitrust suit ever had been brought against a patent contract.[2] Such arrangements were, at the time, a normal and necessary part of doing business. Rohm and Haas owed much of
its success to the exploitation of acquired, patented technology.
In the case of acrylics the cross-licensing was, if anything, even
more crucial to success. All of the named co-conspirators had
research programs in acrylic chemistry and had acquired patents. Without cooperation, it is doubtful whether anyone could
have marketed acrylic products without being dragged into
court on a charge of patent infringement.

The several cross-licensing agreements Haas had signed
with Du Pont fell into the category of such cooperative arrangements. Du Pont had a longstanding agreement with ICI
whereby each would automatically obtain exclusive licenses
for its home country under the patents of the other. Under this
agreement ICI assigned to Du Pont several patents and patent
applications in the area of acrylics during the 1930s. Seven of
these patents and applications conflicted with applications
Rohm and Haas had filed. Both Rohm and Haas and Du Pont
had claimed substantially the same inventions. Establishing
priority so a patent could be issued might take years. Further,
Du Pont already had been issued an American patent on polymethyl methacrylate. Although Rohm and Haas believed no
court would hold this patent valid, ensuing litigation could out-
last the patience of the parties. Until and unless the patent was
overturned in court, it would have been illegal for Rohm and

Haas to use polymethyl methacrylate for any purpose (such as making Plexiglas) without Du Pont permission.

Rather than fight in court, Rohm and Haas and Du Pont signed an agreement to settle these differences on March 5, 1936. They granted each other royalty-free licenses on the disputed patents and applications and arranged to decide jointly on the priority of each conflicting application. With this agreement in effect, Haas was able to begin production and sales of Plexiglas sheet.

The firms signed two further agreements in 1939. The first concerned methacrylic materials for dentures. Both companies had been serving this market for several years when Du Pont received two patents that covered such dentures. Du Pont was willing to license Rohm and Haas, if Haas would agree to Du Pont's price schedule, restrict his sales to the two dental supply houses he already served and sell the material only as preformed, one-ounce, single-denture blanks. Haas reluctantly agreed to the terms after his attorneys advised him that Du Pont's demands were well within its patent rights. Three years later Du Pont agreed to lift these restrictions.

Dentures accounted for only a small part of Rohm and Haas's methacrylic sales, even at the inflated prices mandated by the 1939 license. However, denture prices were of greater public interest than airplane canopy prices, and thus a move to fight price-fixing in dentures was calculated to gain public approval. The Justice Department claimed that the denture wearer was paying fifty dollars per pound for methacrylate polymer which sold industrially at eighty-five cents. This conveniently ignored the profits taken by two sets of intermediaries and the dentists, as well as the time and skill required to make a fitted denture from a blank. Still, the contract price at which Rohm and Haas supplied its two customers amounted to thirteen dollars per pound, a high price even allowing for the costs involved in packaging, the care required to maintain proper color, and the percentage of returns necessitated by the short shelf life of the blanks.

Under the second 1939 agreement, Haas gave Du Pont a license to cast sheet under the Röhm and Bauer patent, but restricted Du Pont's capacity to one-half that of Rohm and Haas. Haas was afraid that giving an unrestricted license to a company whose profits alone were five times his sales might force him out of business. He was able to get Du Pont's agreement to this restriction because Du Pont, notwithstanding the patent, had begun using the sheet casting process in early 1939, after learning about it through ICI. In January 1941 Haas waived the quantity restriction for the duration of the national emergency.[3] Because neither Rohm and Haas nor Du Pont granted acrylic patent licenses to other companies, the two firms divided this new area of chemical technology. Acrylics could be considered a shared monopoly, but both companies agreed it was one legally grounded in valid patents.[4]

What Haas and Du Pont could not have known while negotiating these agreements was that nearly half a century after enactment of the Sherman Act the Justice Department's interpretation of it would change. The change was instituted by Thurman Arnold when he became assistant attorney general in charge of the Antitrust Division in March 1938. Arnold believed the act forbade virtually any type of arrangement that led to a decrease in price competition. President Franklin Roosevelt approved Arnold's appointment after his advisors convinced him a stricter interpretation of the act was desirable social and economic policy. Previous enforcement policy, such as the classic 1911 case that forced the breakup of Standard Oil, had focused narrowly on combines, where monopoly had been gained by the joining of competing firms into a single company.

There were two types of arrangements Arnold found particularly objectionable: patent pools and international agreements with foreign cartels. The latter, by dividing world markets, reduced the number of domestic competitors and decreased opportunities for America's exports. As Arnold could not investigate every possible infringement, he concentrated on those areas where he felt action would produce the most public good. Du Pont seems to have been a particular target of Arnold; he named it as a defendant in at least nine different actions between 1939 and 1942. Arnold brought almost as many court actions in that brief span as all of his predecessors had in the half-century before. Rohm and Haas's acrylic agreements thus attracted Arnold for three reasons—they were based on patents, they involved large foreign cartels, and they included Du Pont.[5]

Arnold's strenuous enforcement of antitrust law soon ran into opposition both within and outside the administration. After December 7, 1941, the nation was at war. All other industrial policies were less important than increasing the production of war materiel. Maximum output required just the sort of close coordination between companies that the Sherman Act sought to prevent. Otto Haas and many others argued further that it was folly to haul executives of companies aiding the war effort into court. Over the course of 1942 and 1943, President Roosevelt came to accept this logic (which had been presented to him by the War and Navy Departments). Roosevelt had most of Arnold's cases either deactivated for the duration, like the Rohm and Haas indictments, or abandoned altogether. Arnold himself left the Justice Department in 1943 to accept a federal judgeship.

In 1944, as war production began to decline, Roosevelt gave the Justice Department permission to reactivate some of the cases. In September the acrylic case was reopened and on December 11, 1944, the Justice Department applied to the Federal District Court in Newark for a trial date. The trial began with the selection of a jury on May 14, 1945, six days after the war in Europe ended. It was the first criminal case against an alleged international cartel to be tried under the Sherman Act. Only the general acrylic indictment was brought to trial, but the government incorporated the accusations from the denture indictment into its case.[6]

The trial lasted for five-and-a-half weeks. These were harrowing weeks for seventy-three-year-old Otto Haas, perhaps the most painful period in his entire life. He sat in the courtroom like a common criminal, facing (at least in his mind) a possible jail sentence. In a detailed repetition of the presentation Hutchinson had made to the congressional committee three years earlier, the government of Haas's adopted land accused him of knowingly participating in a conspiracy that was not only criminal, but detrimental to the national interest of the United States. In a sense, his entire life's work was on trial; he had built his company on adroit exploitation of acquired patents.

Taking the stand in his own defense, Haas described himself to the jury as an immigrant who had decided very early he wanted to be an American and who, as an ethical businessman, believed his moral commitments to Dr. Röhm were as important as legal obligations. He told the jury that, in the contracts which the government alleged amounted to conspiracy, he had sought only to obtain, through legal use of patents, what was necessary to protect his company's interests against much larger competitors.

The trial ended on the morning of June 20 with the judge's charge to the jury. That night the jury returned its verdict: not guilty. All defendants were acquitted. The government had failed to convince the jurors beyond a reasonable doubt that Haas and the others had engaged in a criminal conspiracy.

One eyewitness at the trial, an attorney who was also a personal friend of Otto Haas, expressed the view many years later that it was Haas's testimony that swayed the jury: "I was there at the time and the reason, I think, [for the verdict] . . . was that the jury decided this guy [Haas] is honest and he won't do anything. I think the case was won because of the impression he made."[7]

The Justice Department was, naturally, disappointed. The government attorneys thought they had had a strong case, one they had deliberately selected as the first indictment alleging international conspiracy and patent abuses to be brought to criminal trial. Some department officials speculated the case might have been too intricate for the jury to understand; perhaps they should have brought a civil action instead, as the law allowed and as had been done in many earlier Sherman Act cases.[8]

The acquittal on the criminal charges did not prevent the Justice Department from bringing a civil suit covering the same alleged activities. By October, Wendell Berge, the assistant attorney general for antitrust, had decided to do just that, but to handle Rohm and Haas separately from Du Pont. He thought the government should have won the criminal case, and he knew a civil suit would require a lesser demonstration of proof, merely a "preponderance of evidence."

Going directly into court is not the normal procedure in a civil antitrust action. More often, the government and the prospective defendant attempt to negotiate a consent decree, in which the defendant, while not admitting illegal behavior, promises not to commit the alleged offenses in the future and agrees to abide by a set of corrective measures. A case would be litigated only if the negotiations failed.[9]

Haas decided it would be best to negotiate a settlement rather than fight in court. He was weary from the ordeal and publicity of the Newark trial. It was an experience he wanted to avoid repeating. While the jury had cleared him of criminal activity, he recognized a judge in a civil suit might find some culpability and prescribe actions that would injure his business.

The negotiations dragged on for three years before agreement was reached on all the issues. Rohm and Haas canceled its acrylic licensing agreements with the I.G. and Darmstadt while retaining nonexclusive licenses under their patents. It promised not to enter into similar acrylics agreements in the future and not to give either of the German firms control over

Rohm and Haas's business policies. It also promised not to discriminate in price or sales policy on the basis of end use (a remnant of the old denture indictment). Finally, the firm agreed to give nonexclusive royalty-free licenses under all of its extant acrylic patents to any applicant.

The chief dispute concerned how much of its know-how Rohm and Haas would have to divulge to license applicants. Haas had resigned himself to the inclusion of some know-how clause, as he knew the Justice Department would insist upon it. Berge originally sought a clause that would have required the company to divulge its overall knowledge of acrylic processes, including future advances. Haas feared such broad terms would threaten his hard-won technical superiority and seriously damage the business. Rohm and Haas eventually got the department to agree on far narrower language, requiring it to disclose only the specific knowledge needed to work a particular patent. With this Haas decided he had an agreement with which his firm could live and which was the best he could get without a trial.

On November 18, 1948, the government filed the consent decree in Federal District Court in Philadelphia. The court accepted the decree, and the parties signed it that same day. The decree remained in effect for thirty-five years. In December 1983 the government terminated it as no longer representing desirable economic policy. The decree had little effect on Rohm and Haas's development and prosperity. However, it did prevent the company from acquiring the German Röhm and Haas Company when the opportunity arose in 1953.[10] In a less tangible sense, Haas and his company learned something about the limits the government had placed on the manner in which they could exploit science and technology.

Another major problem confronting Otto Haas in the 1940s resulted from the German minority interests in his companies. In 1940 and 1941, President Roosevelt issued a series of Executive Orders freezing assets held by nationals of Germany and German-occupied countries and requiring firms with substantial German ownership to submit to supervision by the Treasury Department. Under these orders, Rohm and Haas was declared an enemy national because of the Röhm trust, as was Resinous Products because of the Albert Company interest. However, Treasury officials concluded that the two companies were not Nazi fronts, and their operations went on uninterrupted. The principal added burden was more paper work. Samuel Perry, the Treasury official assigned to work at Rohm and

Haas, was so impressed by what he found that he left the government in 1943 to work for the company.

In March 1942 the government revived the Office of the Alien Property Custodian that had been active in World War I. In August the APC seized the minority interests in Rohm and Haas and Resinous Products, and in September two APC officials joined the companies' boards of directors.[11] Otherwise the APC did not interfere in the operation of the companies.

The real question, which came up after the war, was what should be done with the seized stock? In November 1945 Haas asked the government to return the Röhm trust stock to him. He pointed out that the trust agreement had not given his late partner title to the stock, but merely a right to the stock's dividends. A 1931 amendment to the trust agreement provided for automatic termination of the trust in the case of any government attempt to seize it. In answering the suit, the APC made a counterclaim against Haas, demanding an additional ten percent of the company (and accrued dividends and interest) on the grounds that Haas really had been entitled to forty, rather than fifty percent of the company back in 1920. The suits were settled on August 26, 1948, with Haas dropping his claim and paying the government a cash settlement in satisfaction of the counterclaim. The government was now free to sell the Röhm trust stock.

Before the stock sale, Resinous Products was merged into Rohm and Haas. The APC recommended the merger on the grounds it could then get a better price for the Albert firm's interest in Resinous Products. Haas agreed. He had merged Charles Lennig and Company into Rohm and Haas at the end of 1947, and he saw advantages in the Resinous Products merger. It would bring operating economies by unifying the two staffs, and it would end the sometimes bitter competition between them. The Rohm and Haas–Resinous Products merger was approved by stockholders of the two firms on September 14, 1948. Each share of Rohm and Haas common stock brought 10 shares of the merged company, and each share of Resinous Products became 13.7 shares of the new issue.[12]

The government placed Röhm's stock, some 25.7 percent of the outstanding shares of the merged company, up for bids on December 14, 1948. A consortium of investment bankers bid $38.54 per share for the stock on January 17, 1949, and then offered it to the public at $41.25 a share. The attorney general had previously decided that the government's interest could best be served by retail distribution of the stock. The govern-

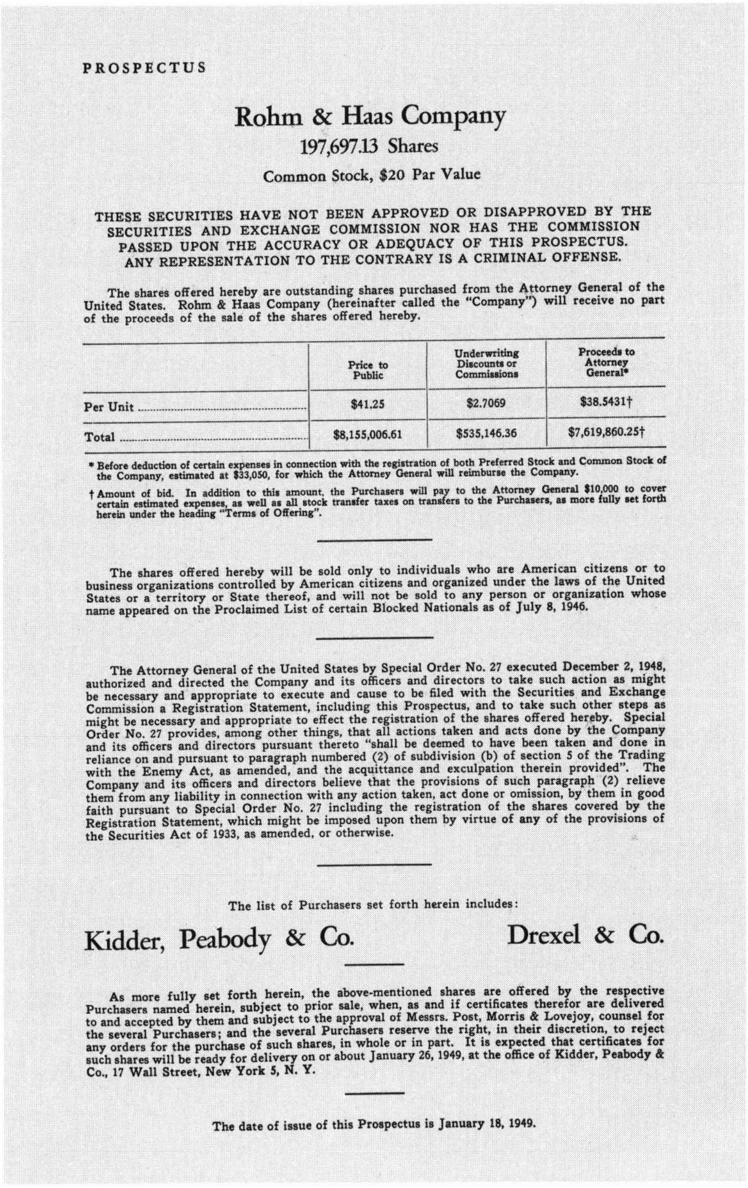

Rohm & Haas Company

197,697.13 Shares

Common Stock, $20 Par Value

THESE SECURITIES HAVE NOT BEEN APPROVED OR DISAPPROVED BY THE SECURITIES AND EXCHANGE COMMISSION NOR HAS THE COMMISSION PASSED UPON THE ACCURACY OR ADEQUACY OF THIS PROSPECTUS. ANY REPRESENTATION TO THE CONTRARY IS A CRIMINAL OFFENSE.

The shares offered hereby are outstanding shares purchased from the Attorney General of the United States. Rohm & Haas Company (hereinafter called the "Company") will receive no part of the proceeds of the sale of the shares offered hereby.

	Price to Public	Underwriting Discounts or Commissions	Proceeds to Attorney General*
Per Unit	$41.25	$2.7069	$38.5431†
Total	$8,155,006.61	$535,146.36	$7,619,860.25†

* Before deduction of certain expenses in connection with the registration of both Preferred Stock and Common Stock of the Company, estimated at $33,050, for which the Attorney General will reimburse the Company.

† Amount of bid. In addition to this amount, the Purchasers will pay to the Attorney General $10,000 to cover certain estimated expenses, as well as all stock transfer taxes on transfers to the Purchasers, as more fully set forth herein under the heading "Terms of Offering".

The shares offered hereby will be sold only to individuals who are American citizens or to business organizations controlled by American citizens and organized under the laws of the United States or a territory or State thereof, and will not be sold to any person or organization whose name appeared on the Proclaimed List of certain Blocked Nationals as of July 8, 1946.

The Attorney General of the United States by Special Order No. 27 executed December 2, 1948, authorized and directed the Company and its officers and directors to take such action as might be necessary and appropriate to execute and cause to be filed with the Securities and Exchange Commission a Registration Statement, including this Prospectus, and to take such other steps as might be necessary and appropriate to effect the registration of the shares offered hereby. Special Order No. 27 provides, among other things, that all actions taken and acts done by the Company and its officers and directors pursuant thereto "shall be deemed to have been taken and done in reliance on and pursuant to paragraph numbered (2) of subdivision (b) of section 5 of the Trading with the Enemy Act, as amended, and the acquittance and exculpation therein provided". The Company and its officers and directors believe that the provisions of such paragraph (2) relieve them from any liability in connection with any action taken, act done or omission, by them in good faith pursuant to Special Order No. 27 including the registration of the shares covered by the Registration Statement, which might be imposed upon them by virtue of any of the provisions of the Securities Act of 1933, as amended, or otherwise.

The list of Purchasers set forth herein includes:

Kidder, Peabody & Co.   Drexel & Co.

As more fully set forth herein, the above-mentioned shares are offered by the respective Purchasers named herein, subject to prior sale, when, as and if certificates therefor are delivered to and accepted by them and subject to the approval of Messrs. Post, Morris & Lovejoy, counsel for the several Purchasers; and the several Purchasers reserve the right, in their discretion, to reject any orders for the purchase of such shares, in whole or in part. It is expected that certificates for such shares will be ready for delivery on or about January 26, 1949, at the office of Kidder, Peabody & Co., 17 Wall Street, New York 5, N. Y.

The date of issue of this Prospectus is January 18, 1949.

Copy of the prospectus issued when Rohm and Haas Company stock was first offered to the public in 1949.

ment disposed of an equivalent block of preferred stock in the same manner. The two blocks of stock brought approximately $9 million to the government. Eventually this money went to Dr. Röhm's children, who convinced the United States they had not been Nazis.[13]

Rohm and Haas became a publicly held company, not as a means of raising capital or giving the founder a chance to "cash in," but because the government had decided it should be one. On March 17, 1949, the Board of Governors of the New York Stock Exchange agreed to Rohm and Haas's application for listing. Trading in the stock began on April 20.

It took the government until June 1956 to dispose of its
other block of Rohm and Haas stock because of a suit by Kurt
Albert's daughter, Irene. An American citizen, she sued for recovery of the stock, contending it had been given or sold to her
before the war. Irene Albert eventually settled for one-ninth of
the disputed stock. Her long-pending suit was actually a favor
to the government. Its remaining stock, some 7.8 percent of the
shares outstanding, fetched $34.4 million; Rohm and Haas
stock had undergone a tenfold price increase since 1949.[14]

By the end of 1948, the legal problems that had plagued
Rohm and Haas for most of the preceding decade had been
resolved. The company had signed a consent decree which
placed certain strictures on the conduct of its acrylics business,
but the limitations seemed likely to cause few real problems.
Haas believed his business was now strong enough and large
enough to stand on its own. Rohm and Haas and Resinous
Products had become a single company, and it had gone public.
While public ownership was a step Haas probably would have
liked to avoid, it was preferable to a competitor owning a size-
able interest. Haas still controlled over half of the stock, and he
remained his own boss.

Throughout the years of legal battles and negotiations with
the government over ownership of Rohm and Haas, Otto Haas
and his subordinates had not neglected the business. They had
been developing strategies for turning the wartime success of
acrylics into peacetime sales and planning ways to participate
in the dynamic, world-dominating economy of the United
States.

Settlement of the legal issues was important in delineating
the relationship between Rohm and Haas and American society and government. But selling chemicals is what Haas would
have preferred to be doing all along. Now, at last, he could turn
his full attention to this.

NOTES

1. *Senate Committee on Patents, Hearing on S. 2303, pt. 2, 77th
Cong., 2nd sess., April 20–25, 1942.*
2. *John Bergin, interview with the author, June 16, 1983, copy in
Rohm and Haas Company archives.*
3. *Senate Patent Hearings, 701, 819, 836, 853.*
4. *Senate Patent Hearings, 843–45.*
5. *Ellis Hawley,* The New Deal and the Problem of Monopoly

(Princeton: Princeton University Press, 1966), pp. 383–471. Simon Whitney, Antitrust Policies: American Experience in Twenty Industries *(New York: 20th Century Fund, 1958), pp. 3–26, 207–25.*

6. *Various articles,* New York Times, *May 15–June 22, 1945.*

7. *Bill Kohler, interview with the author, July 1, 1983, copy in Rohm and Haas Company archives.*

8. *"Course Undecided in Cartel Verdict,"* New York Times, *June 22, 1945. Moffatt to Haas, June 22, 1945, Rohm and Haas Company archives.*

9. *Whitney, pp. 18–19.*

10. *Bergin interview. United States of America vs. Rohm and Haas, Civil No. 9068, E. D. Pa., 1948.*

11. *Memorandum, September 9, 1942, Rohm and Haas Company archives.*

12. *"Rohm and Haas Company and Resinous Products Company: Merger, 1948," Rohm and Haas Company archives.*

13. *Rohm and Haas Company, "Annual Report for 1948," Rohm and Haas Company archives. Kidder Peabody & Company, "Rohm and Haas Company," Prospectus, January 18, 1949. Kohler interview.*

14. *"U.S. Steps Out,"* Business Week, *May 19, 1956, pp. 160–61.*

CHAPTER 7

BUSINESS IN POSTWAR AMERICA

August 14, 1945: Americans literally danced in the streets. It was the day of Japan's surrender, ending World War II. But at Rohm and Haas, the day was significant as much for the volume of canceled Plexiglas orders received as for the celebration. Otto Haas and his newly expanded executive corps were faced with an overriding problem: What would they do with all that capacity for the production of acrylic monomers and Plexiglas sheet? The commitment of company resources to acrylics was so great that if Haas solved the riddle, the company would prosper; if he failed, the continued existence of the company would be in doubt. Growth had been so rapid, and the resources of Rohm and Haas so strained, little time had been available for postwar planning until well into 1945.[1]

Plexiglas sheet sales plummeted as the year progressed. By November they had declined by seventy-five percent from their early-year level. In part, this was because Haas had abandoned fabrication as soon as it was no longer needed for the war; he preferred to sell sheet to customers who would make their own finished goods. Still, it looked as if the company was in for a rough time. The Knoxville plant shut down in November, a victim of falling sales. The plant belonged to the government's Defense Plant Corporation. The government wanted to sell it, but with demand for Plexiglas tumbling, Haas was not sure whether it was wise to buy the plant and thereby increase his commitment to sheet.

Then something unexpected happened: Plexiglas sheet started selling again. Orders began pouring into Washington Square. By March 1946, sales reached eighty percent of the

March 1945 level. Aircraft makers used Plexiglas sheet in the manufacture of small civilian airplanes for what they hoped would be an expanding market. Veterans who had become familiar with Plexiglas during the war used it in starting new businesses. Soon there were Plexiglas cigarette boxes, doll furniture, umbrella handles, giftware, and jewelry. Rohm and Haas boosted this market by assembling a three-room Plexiglas Dream Suite, which it exhibited at department stores around the country. The suite demonstrated home uses of Plexiglas, ranging from edge-lit decorative mirrors to revolving acrylic hatracks.

Because of the surprising revival of sheet sales (and the fear a competitor might emerge), Haas and Vice President Donald Frederick changed their minds about the Knoxville plant. They purchased it from the government early in 1946, reassembled Rohm and Haas's Tennessee workforce and reopened the facility. Through the months that followed, orders for sheet required all that Bristol and Knoxville could produce. The volume of sheet sold in 1946 actually matched that of 1945.

The demand for Plexiglas proved short-lived. After October 1946 sheet sales died as quickly as they had revived. Consumers tired of Plexiglas novelties. The goods they really wanted, made from materials that had all but disappeared during the war years, were returning to the marketplace. A woman who could get gold and silver jewelry had little use for that made of acrylics. The predicted boom in civilian aircraft had not occurred. In February 1947 Haas shut down the Knoxville plant for two weeks and reduced Plexiglas production at Bristol to a single shift.

Sheet sales continued at a slow pace throughout 1947, declining over one-third in both dollars and square feet from 1946, while inflation was running at a double-digit pace. The largest single market for Plexiglas sheet in 1947 was jukeboxes. Three manufacturers collectively purchased three-quarters of a million dollars worth for use in music machine covers. Even this outlet collapsed the next year. Rohm and Haas profits for 1947 were down fifty percent from 1946 while the rest of the chemical industry, and the economy as a whole, were booming.

The following year, Frederick attempted to analyze the company's errors in that period. He said: "we . . . spent too much time with small customers with questionable futures when we should have been concentrating our efforts on developing stable industrial outlets."[2] Rohm and Haas had a long record of successful technical sales development in basic industries like

The Plexiglas Dream Suite, an exhibit which showed the many ways Plexiglas could be used in the home, traveled to department stores throughout the United States in 1945 in an attempt to expand the non-military uses for acrylic products.

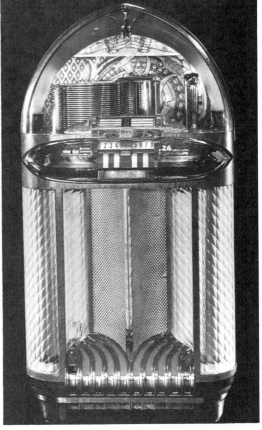

The largest single market for Plexiglas in 1947 consisted of three manufacturers of jukeboxes.

leather, textiles, and (most recently) aircraft, and it was in such outlets that any future success of Plexiglas would be found. Frederick already knew, from limited work done before the war, where these markets might lie: with sign builders, lighting fixture manufacturers, architects, and railroad car designers. These were markets where the product could be sold for its unique performance advantages rather than for its novelty.

Even within acrylics, Frederick had an example of how the right sort of market could be developed. Beginning in 1946, Plexiglas acrylic molding powder, which had been introduced under the trade name Crystalite in 1938, was widely adopted by the resurgent automobile industry. The auto makers used the small polymer pellets to make taillights and reflectors, parts that had previously been made of glass. An early example was the 1946 Buick. Its taillight lenses were made of Plexiglas molding powder colored red. Shortly after the car's October 1945 introduction, Buick began receiving a peculiar complaint: the lens color was fading to a pale amber on one side of the car only, and not always the same side. The explanation was that sunlight beating down on the Buicks while they were parked— and the drivers generally parked them in the same places every workday—was causing the red dye on the taillights to fade in

Plexiglas molding powder has been used to make car taillight lenses since 1945.

Stanton Kelton, Jr., developed coloring techniques for Plexiglas which enabled Rohm and Haas to enter the illuminated sign market.

the sun. Those regularly parked in the shadows were unaffected.

The urgent task of finding a light-stable dye for molding powder fell to Dr. Stanton Kelton, Jr., a well-established member of the Plexiglas production laboratory and the son of the company secretary. Working with a small specialty dye manufacturer, Kelton turned up a dye that not only did not fade on exposure to light but actually darkened slightly. In short order, Plexiglas taillights colored with this new dye became the auto industry standard. Sales of molding powder boomed, increasing some two-and-one-half times to $2.8 million in 1946 alone. The bulk of this total was red molding powder for automobile taillight lenses. Plexiglas molding powder had found a stable use, which has persisted ever since, but one that was tied to the cyclical nature of the automobile industry.[3]

Kelton's new-found knowledge also helped Plexiglas sheet find its first major civilian market in illuminated signs. Before the war there were only two ways to illuminate a sign for night-time use. One was to shine an external light source on it; the other was to outline it with neon tubes. Plexiglas sheet offered a better way. It could be fabricated into a three-dimensional, hollow sign which could be evenly lit from within. To exploit this market, Rohm and Haas needed a wide palette of colored sheet, so customers could replicate their well-known signs in acrylic. Sales representatives in the Plastics Department would sell a company such as Sun Oil Company (an early convert) on Plexiglas signs, and then give the problem of duplicating Sun's yellow and blue colors in translucent Plexiglas to Kelton's laboratory. Kelton and his coworkers proved equal to the task. With signs leading the way, sheet sales began to recover, reaching $10.3 million in 1949 and $12.8 million in 1950.

The market for illuminated, colored signs, such as this early one done for Sun Oil Company, rejuvenated the post-war sales of Plexiglas sheet.

Rohm and Haas encountered an additional obstacle in developing the electric sign market. Plexiglas has a combustibility similar to wood, and some municipal building and fire codes, written before its introduction, were interpreted as precluding its use. Officials in Detroit, for example, refused to issue permits for erection of small Plexiglas signs at General Motors dealers' service centers. In 1948 Rohm and Haas lawyer Frederick "Fritz" Rarig, the company's physics laboratory, and Underwriters Laboratory (the independent testing agency which most municipalities and insurance companies rely on for guidance) helped develop standards for the safe use of Plexiglas, and similar materials, in illuminated signs. In August 1948 Underwriters Laboratory recommended the acceptance of Plexiglas electric signs that met the newly written guidelines. With this report as ammunition, Rarig was able to persuade various municipalities and other standards-writing organizations to rewrite their regulations.[4]

Although signs were the earliest and largest civilian market for sheet, several others developed. Among these were architectural applications such as skylights and room dividers, lighting fixtures and diffusers, and safety glazing. Plexiglas sheet production, however, did not return to levels approaching those of World War II until 1951, when the beginning of the Korean War led to a rebirth of the military aircraft market. Sheet sales in 1951 were over $18 million, a full fifty percent above 1950, on a record volume of 13.4 million square feet. Unlike the World War II period, these totals included no fabrication of finished parts. The company sold sheet and provided its customers with a high level of technical assistance on the manufacture of finished products from Plexiglas. In 1953 Du Pont discontinued acrylic sheet manufacture, leaving Rohm and Haas with no major competition. When Du Pont asked Haas to take over its sheet customers, he discovered his company already had over eighty percent of the market.

Rohm and Haas found other lucrative outlets in the new field of acrylic chemistry. Some products, such as Acryloid industrial coatings and Rhoplex textile finishes, had undergone limited sales development in the 1930s before the demand for sheet cut off most other uses of monomers. One line, Primal leather finishes, already was well established. A number of chemical companies had expressed an interest in purchasing acrylic monomers for copolymer use. Finally, at least one wartime development, Acryloid petroleum additives, was converted to peacetime use with spectacular results.

Rohm and Haas introduced Acryloid 710 viscosity-index im-
prover and Acryloid 150 pour-point depressant for civilian automobile motor oils in 1946. These products brought the same
possibility for extended temperature performance to automo-
biles that the original Acryloid had brought to hydraulic fluids.
They made possible the now-standard multi-grade motor oils,
thereby eliminating the necessity for seasonal motor oil
changes. Rohm and Haas sold almost $1 million of the oil ad-
ditives in 1947, their first full year of production. Sales multiplied rapidly over the next few years, and exceeded $10 million
in 1951, as several oil companies adopted the Acryloids. Oil
additives, a Rohm and Haas discovery derived from the same
acrylic chemistry that had produced Plexiglas, gave Rohm and
Haas entree to a new market and transformed one of that market's staple products.

Oil additives were the most striking success, but Haas's and
Frederick's general faith in the versatility of acrylic chemistry
proved well-founded. By the early 1950s, each of the lines they
had envisioned in the 1940s, except textile finishes, was approaching at least $1 million in sales. The success of these
varied applications was a tribute to the Research Division
which Haas had been building. It had discovered most of the
products and then developed them for specific market niches.

World War II and the devastation of German industry ended
Haas's reliance on purchased technology. Rohm and Haas laboratories, like those of most other American chemical companies, had been growing steadily in both size and stature since
well before the war. In acrylic chemistry, at least, there was no
better laboratory anywhere than the one at Rohm and Haas.
Company scientists delivered a steady stream of new products
under the direction of Dr. Ralph Connor, the research director
Haas had recruited from the government's wartime research
establishment.

One key assignment given the Research Division long before Connor's arrival was to find cheaper ways of manufacturing acrylate and methacrylate monomers. Haas and Frederick
had recognized all along that the cost of acrylic products put a
severe crimp on potential markets, especially in areas such as
coatings and molding powder, where systems based on other,
less expensive classes of monomers provided price competition. Besides seeking less costly sources for starting materials,
company scientists searched for entirely new manufacturing
processes.

One problem involved hydrogen cyanide, a key raw material

In 1946 Rohm and Haas purchased a 534-acre tract of land along the Houston Ship Channel.

for acrylic monomer synthesis. During the war, Rohm and Haas had purchased sodium cyanide from Du Pont to manufacture hydrogen cyanide. This made the company dependent on its chief competitor in acrylic products for a material that was both expensive and in short supply.

Earlier attempts by Rohm and Haas to make hydrogen cyanide had failed, but in 1945 the company tried again. Its Research Division and engineers built a pilot plant at Bristol that manufactured hydrogen cyanide directly through a high temperature catalytic reaction between natural gas and ammonia. The manufacturing process was based on prewar *I. G. Farben* technology, which had been extensively modified by Rohm and Haas scientists. It was both new and unproven. Up to that time, the direct production of hydrogen cyanide never had been accomplished outside the laboratory.[5]

The Bristol pilot facility was a success. Rohm and Haas then decided to build the world's first large plant for direct manufacture of hydrogen cyanide and to locate the facility near a source of natural gas. After a thorough search led by Vice President of Engineering and Construction Lloyd Covert, Rohm and Haas in August 1946 purchased a 534-acre tract fifteen miles southeast of Houston on the Houston ship channel. Three different pipeline companies serviced the area with natural gas, and several petrochemical plants, which could supply other needed starting materials, were located nearby. This purchase was a big step for Haas. It was a step taken with at least a bit of reluctance as it made official something he had known for several years. His company had grown too large for one man to supervise personally. While Haas continued his longstanding weekly trips to the Bridesburg and Bristol plants, he never visited Houston. The new plant site was Rohm and Haas's first voluntary expansion outside the Delaware Valley. The Knoxville plant had been constructed because of wartime concern for the security of coastal regions.

When he purchased the site, Haas planned to spread actual plant construction over several years, so he could finance it out of retained earnings. But in October 1946, Du Pont told Rohm and Haas it would be unable to supply the firm with sodium cyanide beyond December 1947. Construction of the cyanide plant suddenly became a crash program that had to be finished within fifteen months. Covert reported the deadline could be met, barely, if the company took "every reasonable shortcut with respect to procedure in engineering and purchasing."[6] By January, with sheet sales collapsing and construction costs skyrocketing, Haas had to scale down the construction plans. Even so, it remained the largest single construction project the company had undertaken. Haas was increasing the size of his bet on the success of acrylics and, although he had no other choice, it left him more than a bit nervous. As the poor sales of acrylics continued through 1947, Haas postponed other long-planned construction to continue the Houston project.

Despite the top priority Rohm and Haas gave the plant, the late 1947 deadline proved overly optimistic. The company purchased enough cyanide to maintain production until the Houston plant opened in the summer of 1948. The initial installation also included a facility to turn the cyanide, which was too unstable and poisonous for bulk shipment, into acetone cyanohydrin, which could be shipped easily. Acetone cyanohydrin was the next intermediate in methacrylate synthesis. A unit to make the corresponding acrylate intermediate followed the next year. Because of its proximity to natural gas and other petrochemical plants, the Houston plant soon became the company's preferred site for the manufacture of intermediate products for internal use. The 1951 H Process plant for the higher alcohols used in the Acryloid oil additives was the first such installation.

The search for new monomer syntheses led in different directions with the two classes of acrylic monomers. Five years of research failed to uncover a methacrylate synthesis less expensive or otherwise preferable to the process that Britain's Imperial Chemical Industries had developed in the early 1930s. Haas finally had acquired a license for this in 1944 from Du Pont, which owned the American rights. The ICI process was an improved, shorter variation on Dr. Röhm's original commercial technique. The company soon named this variant the B Process. Because cyanide production and sheet marketing had higher priorities, the company continued to use Röhm's original technique (the A Process) until 1952 when the first Rohm and Haas B Process plant opened in Bristol. A second unit followed in Houston in 1955. Although fine-tuned and run in ever larger facilities, B Process chemistry has remained the company's route to methacrylate monomers ever since.

Investigation of alternate acrylate syntheses began in 1946 under Dr. Harry Neher and eventually proved more fruitful. Neher investigated several promising new routes, but kept coming back to one unlikely but intriguing candidate, a reaction discovered by *I. G. Farben*'s Dr. Walter Reppe shortly before World War II. Reppe's process used substantially different starting materials, primarily acetylene and nickel carbonyl. It was plagued by poor yields, slow rates, and excessive by-products. In a 1946 interview, Reppe said he didn't think it had commercial applicability. By 1949 Neher and his associates had proven Reppe wrong. They developed a much modified semi-catalytic synthesis based on Reppe's discovery, which became known at Rohm and Haas as the F Process.[7]

There was one thing standing in the way of construction of an F Process plant: Rohm and Haas's product line used relatively small quantities of acrylate monomers. The three largest acrylic product lines (oil additives, molding powder, and sheet) were all methacrylate based, although some grades of the latter two contained a small percentage of acrylates as softening agents. The chief acrylate polymer lines, specialty emulsions such as Primal leather finishes and Rhoplex textile finishes, had relatively small market potential. An F Process plant would

be economical and lead to decreased monomer costs only if the
company could develop a new, large-scale outlet for acrylate
monomers.

In 1951 Haas and Connor identified house paints as one such
potential application. The use of aqueous acrylic emulsions in
house paints had been suggested about four years earlier by two
young Rohm and Haas emulsion chemists, Benjamin Kine and
Gerald Brown. But the proposal had been rejected, and the scientists had been told to continue working on textile emulsions
instead. Now, however, paint emulsions seemed to Haas a
bright prospect indeed; the total house paint (or trade sales)
market was enormous. He authorized construction of an F
Process plant in Houston and a crash program for development
of an acrylic paint emulsion. The plant came on line in 1952,
and the emulsion, Rhoplex AC-33, appeared the following year.[8]
AC-33 was a product in Haas's favorite mold; it was not a product for consumers, but rather advanced chemical technology
for sale to manufacturers for use in the compounding of water-
based latex paint.

There was at least one other reason why the paint emulsion
project had been approved in 1951 and not when originally
proposed—the 1948 Rohm and Haas merger with Resinous
Products. Paint emulsions required the former's knowledge of
acrylics and the latter's knowledge (especially for marketing)
in coatings. Before 1948 the two firms' employees had been
discouraged from working with each other, and Resinous Products's chemists had been kept away from acrylic chemistry.
Development of an acrylic paint emulsion would have increased friction between the two companies and would have
led to arguments over who should take control of the product.
After the merger there were simply Rohm and Haas scientists
developing a new product in cooperation with and for sale by
the company's Resinous Products Division.

The team of researchers who worked on the paint emulsions
was able to produce a commercial product in under two years.
From their work on textile applications, they were familiar
with acrylic emulsion synthesis and characteristics.[9] They
could measure their work against the new latex-based paints
which had been developed as a by-product of World War II synthetic rubber research.

Both the styrene-butadiene and the polyvinyl acetate latex
emulsions suffered from several major technical problems
which prevented their completely displacing the oil-based systems, but they garnered a lot of attention and respectable sales,

especially in interior paints for the do-it-yourself market. They were easy to use, quick-drying, free of noxious odor, and offered easy clean-up. Rohm and Haas's AC-33 was a major technical advance on these early formulas. It had the durability of acrylic polymers in general and was also far easier to formulate into a finished paint, requiring few of the additives necessary to other latex emulsion formulations.[10]

Despite these advantages, AC-33 sales grew at a modest rate, increasing gradually from $1.4 million in 1954 to $6.1 million in 1958. While paints made from AC-33 were fully competitive with other aqueous emulsions, they too suffered from many technical deficiencies when compared with the oil-based alkyd paints which dominated the market. The acrylic paints had poor leveling and flow rates, which led to rough coats with noticeable brush marks. Low gloss levels and poor adhesion created additional problems; the latter virtually prevented AC-33's use in the formulation of paints for use on wood. Initially, Rohm and Haas scientists tried to overcome these problems through development of modifiers and improved formulations. In the late 1950s, the scientists undertook more fundamental research to solve the deficiencies, but success would not come for some time. When skepticism about AC-33's weatherability in outside use hurt sales, Rohm and Haas began a test program on houses around the country. Connor's own home in a Philadelphia suburb was one of the first houses painted. The success of these tests led to several million dollars of sales for exterior paints in 1956. These sales were for paints for masonry walls, where oil paints will not work at all and the only competition was whitewash.[11]

Cost was still another problem, although perhaps not as great a one as the various technical deficiencies. Because acrylics were more expensive than competing monomers, acrylic latex paints were more costly than both other latex emulsions and traditional oil paints. Of the 320 million gallons of house paint manufactured in America in 1958, only two percent were acrylic latex, while eighteen percent were other water-based systems.[12] Although Rhoplex AC-33 had become an established line by the late 1950s, it was neither one of Rohm and Haas's largest products, nor a major factor in its market. For these things to happen, the company would have to develop emulsions whose technical qualities were good enough to convince paint manufacturers that the advantages in drying time and clean-up outweighed any remaining deficiencies.

Haas could afford to be patient, for his company was prosperous through the 1950s. Having solved the riddle of what to do with acrylics, it enjoyed continued growth. Sales reached $100 million in 1952 and then doubled again before the decade's end.

Haas had succeeded where many other entrepreneurs had failed. He had overseen the transformation of his company from a small firm based on one group of products into a much larger one grounded in a completely different area. The small manufacturer of specialties for the leather and textile trades was now the fountainhead of acrylic plastics. His company, which once relied on acquired German technology, had by 1953 a 600-person Research Division, which included 150 holders of the Ph.D. degree.[13] This staff had proven very adept at working out the commercial implications of acrylic chemistry.

Although acrylic chemistry was now the basis for the largest share of Rohm and Haas's sales, Haas and his staff had not ignored their other lines of business. The company continued to sell Oropon and Leukanol, and developed broader lines of chemicals and finishes for its traditional customers in the leather and textile trades. The Triton surfactants, invented by Herman Bruson in the 1930s, found applications ranging from industrial cleansers to crop-wetting agents. The old Resinous Products lines—resins for the coating industry such as Amberol for varnishes and Uformite and Acryloid for industrial coatings—grew under the direction of Louis Klein, vice president of the Resinous Products Division.

This division developed several new lines as well. Sales of Paraplex plasticizers reached $15 million a year. In a pioneering effort, the division developed ion exchange resins which were sold under the Amberlite trade name chiefly for the purification of water. But the nonacrylic area that saw the most effort, change, and growth well may have been agricultural chemicals.

Lethane, Rohm and Haas's pioneer organic insecticide, had become an important product during World War II but after the war it virtually vanished. Between 1944 and 1948 Lethane sales decreased by ninety-four percent. Lethane was done in by DDT (dichloro-diphenyl-trichloroethane), the world's first modern broad-spectrum insecticide.

DDT was an old compound, but its effectiveness as an insecticide was discovered in 1939 by Dr. Paul Miller of Switzerland's J. R. Geigy Company. Word of his discovery did not reach the United States until 1942. Through a researcher, Dr. Eric Meitzner, Rohm and Haas confirmed DDT's potency but was unable to obtain a license to manufacture it from the War Production Board. Wartime production remained under the strict control of the military, which used the insecticide with heavily publicized success to combat insect-spread diseases such as typhus and malaria. When the war ended, and the government released DDT for civilian use, it swept American agriculture, replacing almost all other insecticides virtually overnight.[14]

Rohm and Haas quickly found that its customers wanted DDT and not Lethane. So it sold them DDT. Initially it repackaged material purchased from a major producer but soon, realizing that this could not be profitable, constructed a small DDT plant at Bristol. DDT had seemed like a sure thing to many other companies, and they all built DDT plants. By 1947, the country was tremendously oversupplied with DDT manufacturing capacity, despite the insecticide's great popularity. Only the largest, most efficient producers could hope to make a profit; Rohm and Haas was not among these. DDT remained a money-loser for the company until it was abandoned in 1955. Haas had permitted himself to be swayed from his basic strategy of selling advanced chemical specialties, and it just did not work.[15]

DDT was not, however, a total loss for the company. Some of the Triton surfactants proved popular as emulsifiers and wetting agents in DDT field sprays. Rhothane, a close relative of DDT, developed and patented by Meitzner, proved superior to DDT for a few specialized uses such as combatting tobacco hookworm. It remained a small but profitable product for twenty years, when it had to be discontinued as part of the 1972 general ban on DDT in America. Two other Rohm and Haas specialty pesticides in the 1950s, Kelthane and Perthane, were also the products of DDT-related chemistry.

The real foundation of Rohm and Haas's major involvement in agricultural chemistry lies in another area entirely, fungicides. It had first become interested in fungicides in the early 1930s, but its first products, two inorganic copper salts, were similar to other products of the day. In 1936 a company researcher, Dr. William Hester, began investigating synthetic organic compounds as possible fungicides. An outside consultant, Dr. James Horsfall of the New York State Agricultural Station of Geneva, worked with him. In 1941, after screening hundreds of compounds, Hester found one promising enough to justify testing on crops. It was disodium ethylene bis-dithiocarbamate. The tests confirmed its effectiveness. When stored as a pure solid, however, the experimental compound decomposed. For this reason, Hester developed a liquid solution which the company named Dithane D-14. It proved effective in several commercial tests in 1943 and 1944, the largest being its success against a 1944 outbreak of potato blight in Texas.

Dithane D-14 had an advantage over copper compounds in that it did not harm the foliage. It also had a disadvantage. Its active ingredient was unstable except in solution, and it decomposed rapidly after drying in the fields. Then the company got a lucky break. Someone added zinc sulfate and lime to a spray tank of D-14. The resulting mixture lengthened the period of crop protection. Hester discovered the extended activity resulted from the formation of a stable, insoluble zinc salt of Dithane in the tank. By 1948 the company was marketing this compound as well, under the name Dithane Z-78. Z-78 broadened the applicability of Dithane to include dry crop dusting as well as wet sprays.[16] Continued company research on this class of chemical revealed the manganese salt was even more potent than the zinc salt, and it was introduced in 1954 as Dithane M-22.

Over the succeeding years, Dithane came to replace the older copper fungicides except where price was a primary consideration. Like many Rohm and Haas products, Dithane wasn't the cheapest answer, just the superior one. Its use spread from potatoes to citrus, tomatoes, grapes, and every other crop where fungus was a problem. Dithane was not the first synthetic fungicide; a Du Pont product preceded it by a few years, but Dithane was the one that transformed agricultural fungus control.

With its position as the fungicide of choice, Dithane soon became the company's largest agricultural product. Domestic sales increased annually through the 1950s, exceeding $6 million by the decade's end and accounting for over forty percent of total agricultural chemical sales. The chief limitation of Dithane's growth was the size of the potential market for fungicides, which was tiny compared to an insecticide like DDT. Dithane nevertheless made Rohm and Haas a visible and continuing force in the agricultural chemicals market and can be seen as the foundation upon which the company's agricultural chemical efforts have been built. Lethane had been a pioneering product but only a transient success; Dithane became a staple, a chemical of choice, and has remained a significant Rohm and Haas product ever since.

A comparison of Rohm and Haas's experiences with Dithane and DDT reveals the difference between those things at which Haas and his company were adept, and those things at

Dithane D-14 was Rohm and Haas's first successful synthetic organic fungicide.

which they were not. Dithane was a unique product, available only from Rohm and Haas, which could be and was sold on the basis of its clear superiority to the competition. By contrast, DDT quickly became a standardized commodity, purchased from any of several suppliers on the basis of price. Potential users already were sold on using DDT. They just wanted to do so at as low a cost as possible, and here Rohm and Haas was unable to compete.

The fifteen years following World War II marked a golden period for American industry, and Rohm and Haas was a full participant. The beginning of the era found Haas and his company in the midst of a fundamental transformation with an uncertain future. The end found the company growing steadily on the solid foundation of its own, now mature, Research Division. It had become, depending on whom you asked, either the largest of the small chemical companies, or the smallest of the large chemical companies, with sales approximately one-tenth of Du Pont, the industry leader. Through it all, Haas, even as he moved well into his eighties, kept Rohm and Haas's activities fruitfully attuned to the way he had always done business: finding advanced specialty products a technically trained sales force could use to help customers produce better finished goods.

Rohm and Haas, like many American companies of the time, found another opportunity for growth in the postwar years. Much of the industrial capacity of Europe lay in rubble, and Europeans looked to America for the material needed to rebuild their economies. Rohm and Haas, although not one of

The stream of new products from the Research Division helped Rohm and Haas participate fully in the golden era of American industry after World War II.

America's giants, found itself called upon to play a part, and this too raised both prospects and problems. How well would Haas's tune play on a foreign stage?

NOTES

1. *Frederick interview. Kelton interview.*

2. *Minutes of the Board of Directors, Rohm and Haas Company, February 26, 1947, Rohm and Haas Company archives.*

3. *Kelton interview.*

4. *"Underwriters Test and Approve Plexiglas,"* Rohm and Haas Reporter, *November 1948, pp. 14–15. Frederick Rarig, conversation with the author, August 18, 1983.*

5. *Leroy Spence, various A Reports 1938–46, Records of the Research Division, Rohm and Haas Company, Spring House, Pennsylvania.*

6. *Minutes of the Board of Directors, Rohm and Haas Company, October 4, 1946, Rohm and Haas Company archives.*

7. *Harry Neher, various A and B Reports 1946–49, Records of the Research Division, Rohm and Haas Company, Spring House, Pennsylvania. Edward Riddle,* Monomeric Acrylic Esters *(New York: Reinhold Publishing, 1954), pp. 1–5.*

8. *Ralph Connor, interview with the author, September 29, 1983, copy in Rohm and Haas Company archives.*

9. *Benjamin Kine, conversation with the author, April 30, 1984.*

10. *Benjamin Kine, "Report 34–36," June 3, 1953, Rohm and Haas Company archives.*

11. *Richard Harren, conversation with the author, August 9, 1984. Connor interview.*

12. *Charles Martens,* Emulsion and Water-Soluble Paints and Coatings *(New York: Reinhold Publishing, 1964), pp. 1–7.*

13. *Ralph Connor, "Report on the Research Division," Minutes of the Board of Directors, Rohm and Haas Company, January 27, 1953, Rohm and Haas Company archives.*

14. *Eric Meitzner, interview with the author, August 16, 1984, copy in Rohm and Haas Company archives. Thomas Dunlap,* DDT: Scientists, Citizens, and Public Policy *(Princeton: Princeton University Press, 1981), pp. 59–74.*

15. *Donald Murphy, "Brief History of the A&SC Department," August 1948, Rohm and Haas Company archives.*

16. *William Hester, various A Reports 1936–46, Records of the Research Division, Rohm and Haas Company, Spring House, Pennsylvania. Murphy, "A&SC."*

FOREIGN OPERATIONS

Otto Haas built his business on domestic sales. For many years his company's foreign operations amounted to little. Rohm and Haas had extensive markets in Canada but, like many American businessmen, Haas treated Canada as part of his normal domestic operations.

Haas deliberately stayed out of Europe. He had no desire to compete with his friend and erstwhile partner, Otto Röhm. The men had agreed, informally in 1920, and formally in 1927, not to compete with each other in products (like Oropon) they both made. Each also agreed to offer the other rights to fruits of his research for sale in their respective home markets. Besides, Haas saw nothing to be gained in competing directly with *I. G. Farben*. The I.G., along with other German firms, had become a major source of new products for Haas's domestic market, an arrangement Haas did not want to disturb.

In the 1930s, however, a change occurred in Haas's attitude towards foreign markets. For the first time, his research unit began developing its own specialties, such as Lethane. After Röhm declined the European rights to Lethane, Haas decided to try to exploit the market himself. In 1936 he opened a small subsidiary in London to sell his pioneer synthetic insecticide. He named the new firm Charles Lennig and Co. (G.B.), Ltd., after an earlier acquisition, and hired E. C. B. Kirsopp's brother George to manage it. By the beginning of World War II, George Kirsopp had developed a small but profitable business, selling American-made Lethane in Britain.

Haas also set up sales subsidiaries bearing the Rohm and Haas name in Canada and Argentina that same year. The

Donald Murphy, vice president for Foreign Operations

former handled a broad line of Rohm and Haas products, while
the latter concentrated on Oropon. The small Oropon plant
this subsidiary erected in Buenos Aires in 1939 was Rohm and
Haas's first foreign manufacturing operation. In largely ignor-
ing overseas markets, Rohm and Haas was little different from
the major American chemical companies. Even Du Pont was
primarily concerned with consolidating its hold at home; it
worked with European companies, rather than competing
against them.[1]

World War II changed all this. As America became, in President Roosevelt's words, "the arsenal of democracy," its industries boomed while Europe's collapsed. After the war,
Europeans sought American help in reconstruction of their

economies. United States firms gained entrance to markets in Europe where previously they had played little part. They also gained influence in formerly hotly contested markets in South America.

South America provided the most immediate opportunities for Rohm and Haas. Between 1944 and 1947 export sales increased from $400,000 to $1.5 million. The three largest markets were Colombia, Argentina, and Brazil, with Mexico and Cuba not far behind. The best selling product was that old standby, Lykopon, but Plexiglas sheet was a close second, and virtually all of the company's lines were represented.[2]

As export sales continued to grow during the following years, European markets gradually supplanted South American ones as the most important. Exports grew to $5.5 million in 1951, almost half of which were to Europe. This growth far surpassed expectations of the top executives in Philadelphia. They had not looked for a significant export volume and had not expanded the export staff. In 1946, Donald Murphy, sales director for agricultural and sanitary chemicals, was given added duties as foreign sales director. As the years went on, the new responsibility took more and more of his time, until in 1954 he became head of a newly created Foreign Operations Division.

A new set of problems developed for Rohm and Haas (and the rest of American business) at the end of the 1940s. The importation of massive quantities of American goods produced severe European balance of payment deficits with the United States. Government trade ministries tried to manage this drain on foreign exchange by placing controls and restrictions on dollar imports, which led American companies to investigate European manufacture as a way to protect and expand their market positions.

His European business had become important enough for Haas to consider manufacturing overseas. In the fall of 1949, Haas sent Vice Presidents Donald Frederick and Ralph Connor to London in search of a British partner for a joint manufacturing venture in accord with the requirements of the British Board of Trade. They carried a letter of introduction to officials of Royal Dutch Shell Ltd., who they thought might be able to suggest an appropriate British partner. Much to their surprise, they found Shell Oil itself seriously interested.

After eighteen months of negotiations, Shell and Rohm and Haas reached a preliminary agreement in July 1951 to build a plant adjacent to a London-area Shell refinery. The agreement called for initial manufacture to include four lines: Triton surfactants, Primal leather finishes, Dithane fungicides, and Paraplex plasticizers. Noticeably absent from this list, and largely absent from exports to Europe, were Plexiglas sheet and molding powder. With Imperial Chemical Industries dominating the British trade in methacrylates, and Darmstadt busy reestablishing its Plexiglas line on the continent, European manufacturers were not clamoring for Philadelphia Plexiglas. At its July 1951 meeting, Rohm and Haas's Board of Directors gave its preliminary approval to the joint venture and authorized the Executive Committee to conclude the deal with Shell. At the last possible moment, Haas changed his mind and called off the agreement.

Rohm and Haas participants in the Shell negotiations disagree as to the root cause of the collapse. The explanation given to Shell, the British government, and the Rohm and Haas board in the fall of 1951 was fear of antitrust violations. Shortly after he learned of the proposal at the July board meeting, Assistant Secretary Fritz Rarig (who had been a federal antitrust prosecutor before joining Rohm and Haas) told Otto Haas there was no way the joint company could operate without running afoul of the Sherman Antitrust Act. John Bergin, the veteran company attorney who had helped draft the contract, disagreed. After two of Rohm and Haas's outside lawyers also gave conflicting advice, Haas withdrew from the venture on September 24.[3]

Many years later Bergin expressed the view that Haas merely used the antitrust issue as an excuse for pulling out of the deal. According to Bergin, the real reason for Haas's withdrawal was he knew he would not control the joint venture. Bergin's explanation has much to recommend it. Throughout his business career, being his own boss and maintaining control had always been of paramount importance to Otto Haas.[4]

In 1950, Haas's proposal to acquire the German Röhm and Haas Company, then run by Dr. Röhm's son, Otto Röhm, Jr., drew a cool reception at the Justice Department whose approval was required by the 1948 consent decree. Haas did not pursue it. Three years later the acquisition of Darmstadt again was considered, but formal application and informal appeal failed to gain Justice Department approval. As a result, the only agreement signed by the two firms in this era was an extremely limited 1951 document on jointly used trademarks: Philadelphia agreed not to use the names and prefixes Oropon, Plexi- and Acry- in certain markets including Europe, and to use Philadelphia as part of the company name in export trade.[5]

But Haas and Murphy, the foreign sales director, were still intent on establishing a plant in Europe. Murphy was particularly intrigued by the potential of Dithane Z-78 fungicide in European agriculture. Fungicides in general were more important in European agriculture than American, especially for use on the vast grape crops of France and Italy. Over the preceding several years, Murphy had arranged for field tests for Dithane on various crops in several countries. The results had been so encouraging that he believed a proposal for local production of Dithane to replace fungicides made from imported copper would be welcomed.

Murphy chose France as the site for the venture. The French market for fungicides had great potential, but perhaps more to the point, Rohm and Haas had a valid patent for Dithane in France, something it had been unable to obtain in several other European countries. In April 1952 agreement was reached with an affiliate of Auby, a large French fertilizer manufacturer under which Rohm and Haas was to supply the manufacturing equipment and raw materials, retain ownership of the material through each step of manufacture, and control distribution. The French firm would provide the building at its plant site in Feuchy in northern France, as well as utilities, labor, and supervision. The plant at Feuchy produced its first commercial batch of Dithane on May 17, 1952, in an installation as large as the one at Bridesburg (six million pounds per year). In 1953 sales of French-made Dithane, and a second product, Paraplex plasticizer, amounted to $600,000. The plant continued to operate smoothly for many years.

Murphy next established a Paris subsidiary, Minoc, to supervise distribution, sales, and government relations throughout Europe. Minoc's first director, Frederick Tetzlaff, ran it from Philadelphia. In 1952 Vincent Gregory, a young executive trained in finance, was sent over to supervise day-to-day operations. All activities on the continent were transferred from Lennig to Minoc in 1954.[6]

Office space was hard to find in Paris and, at first, Gregory worked out of his hotel room. His next office was actually a kitchen and pantry within a large apartment at 3 Avenue du President Wilson. As Minoc grew over the next two-and-one-half years, its office expanded through the apartment one room at a time, until the landlord told Gregory to find Minoc a new office.[7]

Minoc was a great and quick success. By 1955 it had returned Rohm and Haas's original investment in dividends

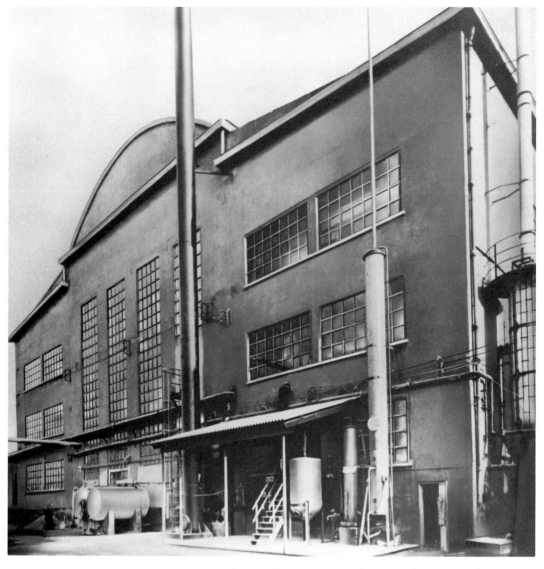

In 1956, Rohm and Haas chartered a manufacturing subsidiary in Treviglio, Italy.

alone. French-made Dithane Z-78 became the established fungicide for a wide variety of crops, including grapes, throughout Europe. Within two years sales of the French-made products reached $3 million. Murphy and his assistants set up a network of independent distributors and supported them with technical service, continued field testing, and coordinated promotion. In addition to the two products manufactured at Feuchy, Minoc served as exclusive continental agent for the full line of imported Rohm and Haas specialties.

The success of Dithane was such that, in 1956, Rohm and

Haas chartered a subsidiary in Italy, Filital, and opened a second European Dithane operation in Treviglio, outside of Milan. At first, Donald Brophy (Minoc's sales supervisor in 1952 and general manager in 1955) commuted regularly between Paris and Treviglio to run Filital. Within a year, the new subsidiary had progressed to the point where it required a full-time resident manager, so Murphy sent Samuel Talucci, a young entomologist and plant pathologist, to Italy.

Meanwhile, Haas and Murphy found a new role for Lennig. They converted it into a manufacturing company. Sales of several products had been limited by British regulations. Amberlite ion exchange resins were threatened by a British law that required Rohm and Haas to manufacture Amberlites in Britain to protect its patent. Tetzlaff located a suitable plant site in Jarrow, near Newcastle-on-Tyne in northern England, and Lennig had a plant up and operating by July 1955. In addition to Amberlites, the plant made Primal leather finishes and Para-

Minoc built its own plant at Lauterbourg in 1958, to benefit from growth in intra-European trade.

plex plasticizers. Plexol oil additives followed the next year (these were actually the Acryloid additives, with the name changed in line with the agreement with Darmstadt).

Murphy installed a new, higher-powered team in London to supervise the expanded effort. As general manager, he selected a veteran engineer, Arthur Miller, who was well acquainted with the construction and operation of new plants. Gregory transferred from Minoc to be assistant general manager. The two men made the transformed Lennig a success. In 1957 the value of Jarrow manufacture exceeded $2 million. By the end of the decade, Lennig was Rohm and Haas's largest overseas operation.

As Rohm and Haas became more established in Europe, the markets for its products continued to grow. Filital began formulating two other pesticides, Kelthane and Karathane, and then started making acrylic emulsions. Karathane proved particularly important as a supplement to Dithane, as it controlled oidium, a powdery mildew, which succumbed to the older inorganic fungicides but not to Z-78. Jarrow kept adding grades of ion exchange resins and oil additives until it manufactured lines as extensive as those in the United States. In 1958 Minoc began construction of its own plant at Lauterbourg, in Alsace, not far from the German border, a site chosen to take advantage of the expected intra-European trade growth which would result from the newly formed Common Market.

Rohm and Haas engineers designed this plant with additional expansion in mind. The initial plans only called for ion exchange resin manufacture, but the infrastructure could accommodate larger, more diverse operations. Lines to manufacture an intermediate for Dithane and for polymerizing and blending oil additives out of Jarrow-made monomers followed shortly.[8]

By the end of the 1950s, Rohm and Haas's sales in Western Europe reached $13 million a year. The business philosophy that had guided Otto Haas in America for nearly half a century worked in Europe as well, but there were differences brought on by the special conditions overseas. Product selection was the most obvious. Plexiglas sheet and molding powder were of negligible importance; Rohm and Haas's methacrylates had no great technical edge over Darmstadt's or ICI's, and they never were pushed. Agricultural chemicals accounted for the largest percentage of European sales.

In South America Rohm and Haas's record was mixed. Its prewar Argentine subsidiary showed little growth because of

The Filibra plant was opened in Brazil in 1955.

government restrictions there. Brazil's policies were more favorable to United States investment. In 1953 Rohm and Haas set up Filibra, a subsidiary in Sao Paulo, and in 1956 Filibra began making Primal leather finishes and Paraplex plasticizers at its own small plant. By the end of the decade, however, Dithane (imported from either the United States or Europe) had become Filibra's most important product.

Canadian operations had been transferred to Murphy's domain in 1954 as part of the creation of the Foreign Operations Division. A plant constructed at West Hill, outside of Toronto, began manufacturing Paraplex, Dithane, acrylic emulsions, and coating resins in 1954, after Canadian law had been changed to make it preferable for American firms to manufacture, rather than just sell, north of the border. In Japan, by contrast, government policy encouraged joint ventures with native firms. So Rohm and Haas and several native partners set up Tocil (Tokyo Organic Chemical Industries Limited) in 1956 to manufacture Dithane fungicides and Amberlite ion exchange resins. Rohm and Haas specialties were sold in many additional countries through brokers and other independent agents.

In these ways, Rohm and Haas's foreign business evolved from an export office that brought in a few hundred thousand dollars per year to a $35 million integrated operation with manufacturing facilities in seven countries on four continents. Much of this success was directly attributable to the personal talents, skill, and drive of Donald Murphy. Although Otto Haas maintained overall supervision, he no longer traveled overseas. He gave Murphy more autonomy than was enjoyed by executives on the domestic side of the company. Murphy spent a good portion of every year flying around the world on Rohm and Haas's behalf. He would think nothing of a two-month trip that took him to Mexico, Brazil, and then across the Atlantic on a twenty-four hour propeller plane flight to Britain, France, and Italy before finally returning to Philadelphia. But even with these trips, he could not do more than plan and implement broad strategies. Out of necessity, as much as anything else, Murphy managed the burgeoning operations with an adroit combination of hands-on supervision and delegation of authority to the young executives running the several operations. Murphy (and Haas) could plan broad strategies, but their implementation had to be entrusted to the subsidiary managers.

Like Murphy, these managers had a freedom of action, an almost entrepreneurial authority, unknown to their peers back home. The requirements of multinational operation produced a certain amount of decentralization. As Dr. Donald Felley, the general manager of Minoc from 1958 to 1964, described it, "In those days you were considerably more remote, autonomous, and independent in running an operation like that one in France. . . . If I telephoned Philadelphia, it was usually to talk to Don Murphy and it might be once a month or once every two months. . . . It was a major occasion."[9] The success of men like Felley, Gregory, Brophy, and Talucci showed that it was possible for Rohm and Haas to succeed without Haas's direct, continual involvement. The business strategies Haas had established could be practiced successfully by others. Rohm and Haas, while remaining in its founder's mold, had become far more than just an extension of one man.

NOTES

1. *Mira Wilkins,* The Maturing of Multinational Enterprise: American Business Abroad from 1914 to 1970 *(Cambridge, Mass.: Harvard University Press, 1974), pp. 78–82.*
2. *Rohm and Haas Company, "Annual Report for 1947," Rohm and*

Haas Company archives. Rohm and Haas Company, "Financial Statements Year 1947," Rohm and Haas Company archives.

3. Minutes of the Board of Directors, Rohm and Haas Company, July 24, 1951, October 23, 1951. Rohm and Haas Company archives.

4. Frederick interview. Frederick Rarig, conversation with the author, August 18, 1983. Bergin interview.

5. Memo, D. Frederick to C. C. Campbell, July 13, 1951, Rohm and Haas Company archives. Assistant Attorney General Stanley Barnes to Otto Haas, December 7, 1953, Rohm and Haas Company archives.

6. Donald Murphy, various reports to Otto Haas, 1951–54, Rohm and Haas Company archives.

7. Vincent Gregory, interview with the author, March 23, 1984, copy in Rohm and Haas Company archives.

8. Murphy, various reports to Otto Haas, 1954–59, Rohm and Haas Company archives.

9. Donald Felley, interview with the author, February 24, 1984, copy in Rohm and Haas Company archives.

ROHM AND HAAS AT FIFTY

Rohm and Haas marked its golden anniversary in 1959 with its remarkable eighty-seven-year-old founder still at the helm. Otto Haas could look back with pride on half a century of accomplishment. From a one-man operation, the business started by the immigrant from Stuttgart in 1909 had grown into a multinational specialty chemical company with more than ten thousand employees, annual sales exceeding $200 million, and a large Research Division constantly developing new products.

As the year progressed, Haas's health began to deteriorate. One day in November 1959, he paid his last visit to the office. Even then his thoughts were not on what he had achieved but on what remained to be done. Production Manager Robert Whitesell, who met with Haas on that final afternoon, recalled: "We talked for an hour about what was new in the company, what was coming, and how he was seeing things. . . . So I remember the last hour the 'old man' worked in this company—he was talking about the future."[1]

Otto Haas had one more thing to do: he had to arrange for his succession. In a certain, limited sense, he had been doing so since the mid-1940s, when he began establishing a series of trusts and a charitable foundation to hold much of his stock in the company. These were directed by his wife, the former Phoebe Waterman, and his two sons, Fritz Otto and John.[2]

Both F. Otto, then forty-four, and John, three years younger, had been employed steadily by the company since the end of World War II. From their youths, their father had directed their development toward expected roles in Rohm and Haas. Both

Otto Haas and his wife, Phoebe, in the yard of their Villanova, Pennsylvania, home.

boys studied chemistry at Amherst College. F. Otto then earned a doctorate in chemistry at Princeton University and John a Master's degree in chemical engineering at the Massachusetts Institute of Technology. F. Otto worked in Leroy Spence's acrylic research lab at Bridesburg before joining the wartime Navy. After the war, he held a series of training positions, first in liaison between research and marketing and later in various sales departments. He also held the title of executive vice president from 1953. As Haas steered his older son toward science and marketing, he steered his younger one toward plants and people. John Haas held training positions in the Bridesburg, Knoxville, and Houston plants before returning to Philadelphia as director of personnel.

Haas asked the board of directors (which consisted largely of senior company executives) to schedule a special meeting for early January to accept his resignation and appoint a new generation of leaders. Only then, in a few individual bedside meetings, did Haas discuss his succession plans with anyone: F. Otto Haas would become president and chief executive officer; Research Director Ralph Connor would become chairman of the board, to counsel F. Otto and give him continued guidance. The board found Haas's choices simultaneously obvious and surprising: obvious because Rohm and Haas was still essentially a family business; surprising because as F. Otto Haas later recalled, "I didn't think [about becoming president] until the very end. . . . I thought he would go on forever."[3] Also, Haas had given no indication of any plans to elevate Connor, who had spent his Rohm and Haas career in research supervision, to a position above the other executives.

On the morning of December 31, 1959, F. Otto Haas received a call from his mother saying his father didn't think he would live until the scheduled meeting date. At a hastily called meeting that afternoon, the board formally accepted Otto Haas's resignation and ratified the appointment of his older son as president and chief executive officer and Connor as board chairman. Otto Haas died in bed two days later, and an era ended.

What did Otto Haas bequeath his successors? He left them a growing and thriving company, with a great, proven strength in acrylic chemistry that brought the company one-half of its sales. There were also smaller but still significant strengths in other areas, including agricultural, textile, and ion exchange chemistry. The greatest legacy, though, may not have been any specific area of applied science or line of products at all, but a philosophy, an implied set of principles which served as a guideline for how Rohm and Haas did business.

This philosophy went back to Haas's earliest days on the road selling Oropon. Haas knew Oropon was a superior product and the way to sell it was to demonstrate its superiority to the individual tanneries. He worked with each tanner to show how Oropon could solve that tanner's problems to their mutual benefit. Haas's efforts never ended with fancy brochures or exhibits at trade shows. He went to potential customers and got his hands dirty. To Haas, the art of selling was the art of customer service.

Not only did Haas develop his method of selling with Oropon, he developed a type of product as well. Oropon came from the cutting edge of science, but it was not a finished product to be sold at retail on its own account. It was a specialty to be sold to a group of manufacturers on the basis of its superior properties for use in the production of their products. Oropon would lose its identity during this processing. A consumer would be very unlikely to select a pair of shoes because their uppers had been bated with Oropon, but he was likely to notice that the leather was of good quality and appearance.

As Haas introduced new products and expanded markets to serve additional industries, he repeated the strategies learned from Oropon. Leukanol, Amberol, Rhoplex AC-33, Plexiglas, and many others were advanced chemical specialties to be used by other industries in the manufacture of their goods. Rohm and Haas products were neither commodities nor consumer goods. New products were sold as Oropon had been sold, by technically skilled salesmen working with industrial customers. The best of these salesmen were the ones who truly understood the product and its potential. Over the years many employees, like Donald Frederick and Donald Murphy, began in technical positions only to find themselves eventually in sales. Time and again, Rohm and Haas found success in this combination of advanced science and technical sales.

The institutionalization of innovation required a constant flow of new science. During this half-century, the main source of innovation gradually evolved from Dr. Röhm, to other purchased German industrial science, to a large, productive in-house research division. The basic strategies remained the same, although the details changed as the company became larger and moved into grander arenas.

Otto Haas was not just a business strategist. As the company grew, he became, as was necessary, a strong leader and supervisor of men and women. His personality dominated the company; more than twenty-five years after his death, Rohm and Haas employees still retell and cherish stories about him. These tales paint a picture of a man who inspired intense loyalties and cared deeply about the welfare of his employees but who, at the same time, was capable of driving people endlessly and showing little tolerance for failure.

As is common among entrepreneurs, Haas was a firm believer in centralized, personal control. He wanted to know everything going on all over his company. Well into the 1930s (at which time the volume finally got too large) he read every piece of incoming mail before it went to the appropriate individual and checked every bill before it was paid. He did not care much for working through channels or organizational charts. When he wanted to speak to somebody, regardless of where that person was in the corporate hierarchy, he spoke to him, and not necessarily on the subject that was officially the employee's responsibility. Donald Murphy was head of Agricultural and Sanitary Chemicals and Foreign Operations, but Haas was as likely to seek his opinion on leather chemicals as on anything else. He knew that the growing company of the 1940s and 1950s required more hierarchy and specialization, and he went along with it. But he always preferred the old way. Haas told many people that he had a lot more fun when he didn't have to deal with any vice presidents.[4]

In his quest for business success, Haas put in long hours and expected the same of everyone else. If any employee dared to protest he was doing the best he could, Haas would likely look out his window onto Washington Square, point to one of the derelicts who then frequented the park, and say, "You see that bum there? He's just doing the best he can, too." In the late 1940s, while he struggled to establish civilian markets for Plexiglas sheet, he hounded the marketing people morning, noon, and night. In addition, he required seven reports daily on the marketing effort. As Production Manager Whitesell put it, "Why did we beat Du Pont? They didn't have Mr. Haas driving their sheet business like Mr. Haas drove the sheet business here. . . . Just Drive! Drive! Drive!"[5] This could cause some people to seek employment elsewhere and send others into a shell for fear of incurring Otto Haas's wrath. Nevertheless, his leadership brought out the best in many others. He was often a difficult man to work for, but they knew when the chips were down, he would stand behind them.

Although a formal pension plan was not set up until 1943,

Haas saw to it that any employee who was too old or otherwise unable to work had what was needed to survive. On his death bed, he told Donald Frederick, "One thing I want done. . . . If there are any loyal employees who they, themselves, or people in their family are in need of things they can't handle, I want the company or my personal [funds] to handle it."[6]

Haas was extremely proud Rohm and Haas had made it through the depression without laying off a single worker. If there weren't enough orders to fill, he put his employees to work paving plant roads, painting buildings, and doing other fix-up chores. To brighten his workers' off-duty hours, he built employee clubhouses at both prewar plants and maintained a beach and picnic ground on the Delaware River at Bristol. In return for his efforts, Haas had a devoted workforce. His paternalistic attitude may seem anachronistic and even inappropriate today, but it was characteristic of the Bismarckian Germany of his youth.

Many stories survive of Haas's concern for his employees. After a worker stole the Bridesburg payroll back in the 1920s or 1930s, Haas pressed charges and the man was jailed. Some months later, as F. Otto Haas tells it, "a woebegone lady and a flock of little kids came into the office and said, 'We're hungry.' It was Mrs. D. [the thief's wife]. So my father called up City Hall and said, 'You remember that fellow D.? Well, you've got to get him out of jail.' So pretty soon D. was back out of jail. My father set him up in the trucking business in Florida and apparently he went straight."[7]

The boss's sharp tongue was also legendary. When a young plant engineer suggested a way of improving the production of Lykopon, Haas invested $40,000 in the idea, but it didn't work. For the rest of the engineer's career, the poor fellow was known to Haas as "that fool who lost $40,000 for us."[8] Sometimes, too, Haas's temper could get the best of him. He would fire people, only to hire them back with a raise as soon as he calmed down. This happened to a guard at the Bristol plant gate during World War II. The government had placed strict security measures on the plant because of the vital war material (chiefly Plexiglas) being produced there. When Haas, who spoke with a German accent, arrived at the gate one day, the guard refused to admit him because he didn't have a pass and wasn't on the day's pass list. Haas angrily told the guard, "Don't you know who I am? I own the place. If you don't let me in, you're fired." Finally, the guard called the plant manager who, much embarrassed, let

Otto Haas and employees at a picnic in Bristol in the 1920s.

Haas in. When Haas calmed down, he realized the guard had only been doing his job, and he rehired the man with a five-dollar raise.

Alongside Haas's genuine old-world concern, there was a certain amount of old-world formality. To everyone at the company, from the newest mail clerk to experienced scientists and vice presidents, the head of the company was always referred to as Mr. Haas—never simply Haas or Otto.

What Haas had little time for or interest in was outside appearance and display. The company didn't have flashy offices and avoided things that would bring it public attention. Haas took the train and the streetcar to the office, and thought everyone else should too. He had no use for big cars, and told Whitesell he was a fool for getting one. He hated to have his picture taken; publicity and public aggrandizement were things to be avoided. The portrait of Haas that hangs today in the chairman's office is a small painting derived from a passport photograph. Similarly, a well-known (within the company) picture of Otto Haas walking out of the Washington Square building in 1959 exists only because a photographer who had been hired to photograph the building's front door saw Haas walking through the door and casually clicked the shutter without letting Haas know.

Haas cared little for the investment community's opinion. The government may have forced his company to go public, but it could not change his mode of operations. In the late 1950s, Rohm and Haas stock was the highest-priced stock on the New York Stock Exchange, regularly exceeding $500 per share, despite offering less than $2 in yearly cash dividends. The Exchange implored Haas to split his company's stock and bring it down to a more reasonable trading range, but he refused.

Like many immigrants, Haas was intensely patriotic. In the years prior to World War II, he kept expanding his company's Plexiglas capacity because he believed, correctly as it turned out, the war effort would require far more acrylic sheet than the government expected. From 1949 to 1970 Rohm and Haas operated, at no profit, a rocket propellant research laboratory for the government at the Redstone Arsenal in Huntsville, Alabama, even though this topic was far from any of Rohm and Haas's commercial concerns.*[9]

As Rohm and Haas's fiftieth anniversary year came to a conclusion, a new generation of leadership prepared to take over. But F. Otto Haas and the other executives faced a transformation which in its own way could rival the rise of acrylics in its ramifications. The founder had passed from the scene, and with his passing the entrepreneurial era ended at Rohm and Haas. The management style of the company would have to evolve into something more appropriate to a large, complex firm. The change would have to be accomplished without disturbing the company's underlying strengths.

*In September 1985, the United States government again paid tribute to Mr. Haas's patriotism when it presented him with a posthumous Department of Defense medal for distinguished public service in operating the Huntsville facility.

NOTES

1. *Robert Whitesell, interview with the author, September 19, 1983, copy in Rohm and Haas Company archives.*

2. *Waldemar Nielson,* The Big Foundations *(New York: Columbia University Press, 1972), pp. 239–43.*

3. *F. Otto Haas, interview with the author, April 13, 1984, copy in Rohm and Haas Company archives.*

4. *Whitesell interview.*

5. *Whitesell interview.*

6. *Frederick interview.*

7. *F. O. Haas interview.*

8. *Whitesell interview.*

9. *Connor interview.*

PART THREE ◂

THE MODERN CORPORATION 1960 ⋈ 1984

CHAPTER 10

THE POST-ENTREPRENEURIAL ERA

F. Otto Haas realized that, like many sons of corporate founders, he was ill-prepared to succeed his entrepreneur father. Although he had worked in a cross-section of sales departments and had acquired broad familiarity with the company's products, he never had held a decision-making position, nor had he been part of his father's inner councils. Also, his academic training had been exclusively in science. Sensibly, he decided to give himself the on-the-job training he needed. He settled into his father's routines, making the rounds of the able counselors he had inherited.[1]

Little by little, he began to make changes. He perceived that the highly centralized, informal style to which his father had clung would not work for him, nor for the Rohm and Haas of the 1960s. He concluded the company would function best if he acted more as a team leader than as a dominating presence. He chose to delegate authority to the firm's many able managers and executives, and to reach major decisions by consensus. Delegation of authority required clearly defined job descriptions, and executives began to find their job titles corresponded more closely with what they were being asked to do. No longer was the patent attorney consulted about the advisability of pursuing foreign agreements. The corporate secretary lost his responsibility for developing flammability tests. Rohm and Haas was now clearly bigger than one man.

Simultaneously, F. Otto Haas (hereafter simply Haas) started the long and difficult process of rationalizing the organization. In early 1962, J. Fay Hall became the first general counsel in the company's history. Previously, the heads of the several legal departments (regulatory, patent, taxes, general, and the corporate secretary) had reported independently to the top. Later that year, a broader-based administrative department appeared under the direction of Executive Vice President John Haas. John Haas had been in charge of personnel since 1953, and had supervised the purchasing department since the 1959 death of Purchasing Supervisor P. J. Clarke. In addition, he now had the traffic department reporting to him. More than a change in title or reporting relationship, what John Haas noticed most was an increased authority to make decisions concerning the "people side" of the company. Twenty-five years later he remarked: "Right up to 1960 I felt very much under the thumb of my father because he had to clear almost everything. . . . I don't think I felt I had an awful lot of latitude."[2] Increased authority was enjoyed by executives throughout the firm.

Rohm and Haas's most significant rationalization of 1962 was the elimination of the Resinous Products Division. All of the company's domestic sales and marketing efforts were finally placed in a single structure under Vice President Donald Frederick's supervision. The following year Frederick reorganized the sales departments. Resins and plastics formed one new sales division, and industrial chemicals (which included agricultural chemicals) the other.

A far more extensive reorganization followed in 1965, shortly before Frederick's retirement, when the company took its first steps away from a functional-department structure toward a multidivisional one. This was a move most of its competitors already had taken.[3] Rohm and Haas set up three operating divisions: Chemicals, Foreign Operations (later renamed International), and Fibers. A fourth division, Health Products, followed shortly. The Fibers and Health Products Divisions resulted from the company's efforts at diversification. Other departments, including Legal, Financial, Research, Engineering, and Administration, became staff divisions, operating on a corporate-wide basis. The key Chemicals Division, which contained all of the company's traditional domestic businesses, was placed under the direction of Vice President Robert Whitesell, who formerly had been in charge of production. Rohm and Haas had not gone as far as the textbooks proposed in adopting this structure. Most notably, Whitesell had control of only two facets of the core business operation: production and sales. Other functions crucial to his division's operations, including purchasing and research, remained corporate staff departments.

John Haas directed the company's new administrative department, which included personnel,
purchasing, and traffic.

Robert Whitesell, vice president for the Chemicals Division

*Philadelphia Mayor James Tate and Rohm and Haas President F. Otto
Haas break ground for the new Rohm and Haas headquarters at Independence Mall. The shovel was made from Plexiglas.*

Another matter which received Haas's attention in the early
1960s was the need for a new headquarters building. After out-
growing its Washington Square facility, the company moved
several departments into rented space in a dozen nearby buildings and others to a company-owned office building at 17th and
Walnut streets. Haas considered erecting a new headquarters
on Washington Square or moving to the suburbs. But in 1961
he committed the company to participation in Philadelphia's
urban development plan for the historic Independence Hall
area, a few blocks northeast of Washington Square. The city
agreed to acquire and assemble, via eminent domain, a suitable
site on the southwest corner of 6th and Market streets, which
it would turn over to Rohm and Haas in exchange for its
Washington Square headquarters and other buildings in the vi-

cinity which the founder had acquired during the 1950s. The building Rohm and Haas constructed, from a design by Dean Pietro Belluschi of the Massachusetts Institute of Technology, became the company's new world headquarters on June 26, 1965. Rohm and Haas had the distinction of being a neighbor to the Liberty Bell.

Rohm and Haas's research facilities also were cramped. They had spread beyond the much expanded Bridesburg Building 60 to include several departments at Bristol. Moreover, the company had found its Bridesburg plant location a handicap in the recruitment of scientists. Many competitors with whom Rohm and Haas was vying for talent had built aesthetically pleasing, campuslike research centers in rural and suburban areas. Haas decided his company, too, should have a research campus. In June 1960 he purchased a 140-acre farm in Spring House, Pennsylvania, about twenty-five miles north of Independence Hall. The first three buildings at the Spring House site opened in September 1963. More buildings went up during

The company started consolidating its research activities with the purchase of a 140-acre farm in Spring House, Pennsylvania, and construction of three buildings in 1963.

the following two decades as the Research Division moved, one laboratory at a time, to its new home.

These changes took place against a backdrop of continued business success. Between 1960 and 1965, Rohm and Haas's domestic net earnings climbed from $18.3 million to $28.1 million and its sales from $190 million to $241.8 million. Overall figures (including foreign subsidiaries and exports) showed an even more spectacular rise, but comparison is difficult because data from foreign operations were not consolidated into the company's financial statements until 1962.

During this period, domestic Plexiglas sheet sales rose from $29.7 million to $37.9 million as acrylics increased their share of the sign market and gained greater penetration in safety glazing and other architectural areas. Acryloid oil additives and Plexiglas molding powder sales remained stagnant. They had already captured their major markets, multi-grade motor oils and automotive taillights.[4] Agricultural sales, led by Dithane, held steady domestically and continued to grow overseas as the company's foreign operations expanded.

Even a relatively mature business like acrylics required continued investment. Rohm and Haas had added a second F Process plant for acrylates at Houston in 1958, increasing its acrylate capacity from five million to seven million pounds per month. Construction in the early sixties added capacity for Plexiglas sheet and several starting materials.

In 1960 the federal government decided to dispose of a Louisville, Kentucky, plant that it had built during World War II to make butadiene, a component of synthetic rubber. Rohm and Haas purchased it for $6.1 million in 1960. The company recouped much of its investment by dismantling the obsolete butadiene installations and selling them overseas.[5] Construction at Louisville of a B Process plant for methacrylates was completed by the end of 1961, but was immediately mothballed when demand for methacrylates lagged behind projections. In 1962, adipic acid, a plasticizer intermediate, became the first product produced at Louisville.

The perception that commercial acrylic chemistry had reached a relatively mature state by no means implied a discontinuation of Rohm and Haas's acrylic research, but it did mean a major part of that research was directed more toward improvement of existing processes and products than toward creating new product lines. With much of acrylics no longer on the leading edge of technology, there was an increased risk that a competitor could cut into Rohm and Haas's markets. The best way to forestall such incursions was to make sure Rohm and Haas had the lowest cost processes for the production of both monomers and polymers. Out of this program in 1966 came a continuous extrusion process for Plexiglas molding powder, a premier example of Rohm and Haas research ingenuity in the 1960s.[6] The first continuous molding powder line went up at Louisville in 1966. A second Louisville line and a line at Bristol were built soon after.

Another way in which Rohm and Haas exploited its leading position in acrylic chemistry was through economies of scale. By 1968 Rohm and Haas had four separate plants in the United States for the manufacture of methacrylate monomer, two at Houston and others at Louisville and Bristol, each with a capacity of seventy-two million pounds per year. Further, the company expected that it would need another plant within two years. Instead of building a fifth seventy-two-million-pound facility, Chemicals Division head Robert Whitesell won company approval for a mammoth four-hundred-million-pounds-per-year plant in Houston to replace them all. Although the chemistry of the new plant was still the reliable B Process (and thus basically the same as the older installations), the sheer scale of the plant lowered costs sufficiently to frighten off several potential competitors.[7] This pattern of low-cost, high-volume, mass production had become common in the chemical industry.

The need for additional methyl methacrylate capacity partly resulted from unexpected demand for Rhoplex aqueous emulsions. The first Rhoplex emulsions had been introduced in the late 1930s for use as textile finishes. Like those that followed, these early Rhoplexes were copolymers containing both acrylates and methacrylates; the former contributed flexibility and the latter hardness. The markets for acrylic emulsions expanded with the 1953 introduction of Rhoplex AC-33, the first emulsion for house paint. Over the years the several emulsion markets had grown into stable but fairly small businesses. In 1960 Rohm and Haas sold $8.1 million of Rhoplex in the United States for textile and paper coatings, and $7.6 million for house paint. Five years later, the sales figures were $8.5 million and $10.2 million, and five years after that, in 1970, the totals reached $17.8 million and $24.6 million with no end of the curve in sight. These figures represented real growth; prices had declined by approximately twenty-five percent in the first half of the decade, and remained stable after that. The increase in textile emulsion sales was largely the result of market expansion. Nonwoven fabrics, especially cover stocks for disposable diapers, were made with Rhoplex emulsions and became important consumer products in the late sixties.

The growth of paint emulsions, on the other hand, was largely the result of years of patient research on the fundamental properties of good paint vehicles—research which paid off in a series of improved emulsions more competitive with the dominant oil-based alkyd paints in properties such as flow, gloss, and hardness. The research effort was so long in reaching commercial fruition that in the early sixties Haas and the rest of top management seriously considered abandoning the field altogether.[8] A decade after their introduction, acrylic emulsions were only a minor factor in the paint trade and still showed no sign of growing.

Over the next dozen years, scientists at Rohm and Haas meticulously pieced together the performance characteristics that were necessary to please the American consumer. Paint formulators who relied on the technical prowess of Rohm and Haas were the beneficiaries of a whole new series of Rhoplex formulations introduced to the market by the names: AC-34, AC-22, AC-388, AC-490, AC-507, and AC-64. Each acrylic emulsion improved upon the earlier version. Substantial gains were made in flow and leveling for interior flat paints; in durability, gloss, and hardness for exterior flat paints; and in chalk adhesion. These improvements, when combined with the advantages of acrylics—ease of use and clean-up, and fast drying—produced major sales gains.[9]

By 1973 Rohm and Haas emulsions, led by AC-490, had captured the leading position in the interior semi-gloss market. Exterior flat paint sales, chiefly AC-388, continued to increase for several additional years; by 1978 acrylic emulsions had all but supplanted oil-based exterior paints in the domestic market. In 1975 Rohm and Haas sold more than $58 million of these paint emulsions in the United States alone, and trade sales replaced Plexiglas sheet as Rohm and Haas's top selling domestic product line.

With a few scattered exceptions, Rohm and Haas was unable to duplicate the domestic success of acrylics in overseas markets. In Europe and most of the Pacific Basin, consumers equated paint quality with gloss, and were unwilling to accept the tradeoff which American consumers (who in any case were accustomed to less glossy paints) had welcomed. But Rohm and Haas's growing expertise led to a wide range of acrylic emulsions for other uses—printing inks, floor polishes, industrial (factory-applied) coatings, and maintenance paints for metals.

the great cover up.

Now you can formulate exterior paints with outstanding one-coat repaint coverage. Use RHOPLEX® AC-388 100%-acrylic emulsion. Here's the key to the exceptional opacity of paints made with this vehicle: superb flow, leveling, and film build.

RHOPLEX AC-388-based paints also offer excellent tint retention, resistance to dirt pickup, chalk resistance, stability with universal colorants, and superb overall outdoor durability.

In addition to being a stellar performer in exterior housepaints and gloss trim paints, RHOPLEX AC-388 also shows cost/performance advantages over competitive systems in interior flat and semi-gloss formulations. Write for a sample of RHOPLEX AC-388, technical literature, and an appointment to visit our test fences located near Philadelphia, Pa.

ROHM
and HAAS
PHILADELPHIA, PENNSYLVANIA 19105

A 1969 advertisement for Rhoplex AC-388, the breakthrough product which established Rohm and Haas's acrylic paint emulsions as a major force in the exterior paint market.

Many of these products found equal success overseas and in the United States.

The emulsions were about fifty percent water, making shipping costs high. To meet the increased demand and to reduce the cost of shipping all that water great distances, Rohm and Haas built a series of regional plants. New and larger kettles for preparing the emulsions went in at Louisville, Bristol, Knoxville, and then again at Louisville. A new plant opened in 1969 in Croydon, Pennsylvania, across the road from the Bristol plant. Still another emulsion plant opened in 1972 in Hayward, California, to serve the western market. Haas never flinched from making these investments in emulsion facilities, even though they pushed the capital budget far above planned levels.[10]

The emulsion explosion put pressure on monomer facilities as well. By the end of 1968, the company had three F Process plants running at Houston for acrylates and was running four B Process methacrylate plants at close to capacity. In 1970 the company built a fourth F Process plant at Houston, even though an entirely new route for acrylate synthesis, propylene oxidation or the T Process, had reached a sufficiently advanced stage of development to permit construction of a small plant at a new site in Teesside, England. Rohm and Haas could not afford to wait for the T Process to prove itself. The Teesside works opened in 1972 with a B Process plant for methacrylate monomers in addition to the T Process unit. Numerous problems and bottlenecks appeared in the new technology, and several years elapsed before the plant was operating smoothly at rated capacity.

Paint is a complex substance, formulated of many components in addition to binders (or vehicles) such as the Rhoplex emulsions. The close relationship which developed between coatings research, marketing, and several paint manufacturers led the company into a broad range of products for paint formulation. Some were straightforward applications. Tamol dispersants and Triton surfactants were old Rohm and Haas lines, and the development of Tamol and Triton for paint posed few problems. Other areas required more research. In 1971 Rohm and Haas introduced Skane M-8, the first nonmercurial mildewcide for paint, an additive which was particularly important in warm, humid climates. The company also introduced a line of Acrysol soluble acrylic thickeners. Thickeners are needed in latex paints because aqueous emulsions normally are too thin for satisfactory application.

The success of paint emulsions led Whitesell to further realign the Chemicals Division sales departments in 1968. The Resins and Special Products Departments disappeared and their lines were redistributed among newly created Coatings and Industrial Chemicals Departments. This reorganization wiped out the last vestiges of the old Resinous Products organization, and went far toward assuring that each market segment would be served by a single sales department.

Although the reorganizations and decentralization were necessary at Rohm and Haas, they brought some problems. Decentralization of authority, when combined with Haas's inclination to reach decisions by consensus, made the company more vulnerable to breakdowns in coordination between its component parts. In Whitesell's view at least, this problem existed between his Chemicals Division and the Research Division.[11]

The Research Division, under the direction of Chairman of the Board and Vice President for Research Ralph Connor and Research Director Charles McBurney, by and large set its own priorities. This worked well enough when the question was one of improving existing product lines such as Plexiglas molding powder and acrylic emulsions, but it proved less successful in pioneering research. Several new products came out of research in the sixties, including sludge dispersants, flocculants for sewage treatment, and acrylic films—but without notable success. In later years Whitesell commented he would have preferred to have seen more money spent in areas such as agricultural chemicals where the rewards for success would have been greater.[12]

One reason Whitesell urged an increased commitment to agricultural chemicals was the overseas performance of Rohm and Haas's fungicides. Dithane fungicide continued throughout the decade to be the bellwether of Rohm and Haas agricultural products, even though competition caused prices and margins to decline after the expiration of the basic American patents in 1960 (and the French patents two years later). A new formulation with greater and longer-lasting effectiveness, Dithane M-45 reached the market in 1962 and soon accounted for the largest portion of Dithane sales. M-45 was a coordination product containing both zinc and manganese ions. Dithane remained far more important overseas than domestically. In 1967 for example, the Chemicals Division sold $6.9 million in the United States, while Foreign Operations sold $18.5 million (two-thirds of which was manufactured at plants in France and Italy).[13]

Rohm and Haas introduced two new herbicides in the early sixties, Stam in 1961, and Tok the following year. Stam, an herbicide useful on rice under western (but not oriental) growing conditions, was the first chemical to prove effective in controlling unwanted grasses without harming the rice (itself a grass). Rohm and Haas's development and introduction of Stam was based on the patented use of a previously known substance as an herbicide. The validity of its patent came under legal attacks that persisted for more than twenty years and concluded with the ultimate invalidation of the patent in 1984.

With agricultural chemicals such as these accounting for

around a third of division sales through the 1960s, the Foreign Operations Divison under Donald Murphy continued the pattern of high growth it had begun in the 1950s. Foreign sales, including both exports and foreign manufacture, more than tripled during the decade, increasing from $44.8 million in 1960 to $144 million in 1969, while overall company sales only doubled to $448.2 million. Not only did European operations continue their expansion, but Rohm and Haas also achieved greater penetration of markets throughout the rest of the world. This pattern was then common to many large American companies which, having established themselves in Europe during the postwar years, were now expanding further.

As the standard of living throughout Western Europe rose, demand for sophisticated chemicals such as those Rohm and Haas produced increased. But competition, especially from Germany, got tougher, too. With the rebuilding of its industry, Europe ceased to be an easy market for sales achievement. American companies needed managers with decision-making authority close to the scene, who could react quickly to changing conditions and competition and deal with the increasing complexity of business there. Murphy decided to reorganize Rohm and Haas's European operations. He set up a regional headquarters in London in 1964 under the direction of Vincent Gregory, who previously had been general manager of Lennig.

Lennig, Filital, and Minoc now reported to Gregory, rather than directly to Philadelphia. Many of the decisions affecting Europe were thereafter made in Europe, allowing for shorter lead times and more immediate coordination among the businesses in the several countries.[14] By 1969, Samuel Talucci, who had succeeded Gregory as head of European operations the preceding year, was running a $69.6-million business. Decentralization, which transferred decision making from top management to officials closer to the markets and customers, occurred in Rohm and Haas's foreign operations before it happened in the firm's far larger domestic chemicals business.

The range of products manufactured at each European subsidiary's manufacturing plants continually expanded. By the end of the decade, every plant had a kettle for the manufacture of acrylic emulsions. In addition, Minoc, the French subsidiary, had added lines for Dithane M-22, Dithane M-45, and Triton surfactants at Lauterbourg, and Tok (which French farmers found valuable for controlling weeds in wheat fields) at Feuchy. Lennig in Great Britain became the only European subsidiary with a substantial plastics business, manufacturing acrylate and methacrylate monomers, and acrylic sheet and molding powder at Jarrow, and then additional monomer at Teesside. Lennig sold sheet and molding powder under the name Oroglas, because Rohm and Haas Darmstadt owned the Plexiglas trademark outside of the Americas. Besides adding Tok, a number of leather finishes, and several other products in Treviglio, Filital built a plant of its own at Mozzanica in 1965, which soon began manufacture of herbicides and ion exchange resins. Finally, Rohm and Haas set up a new subsidiary in Spain, Ebro Quimica Comercial S.A., which built a small plant at Tudela, outside of Barcelona. In addition to the manufacturing expansion, both Minoc and Filital established sales agencies in other countries within their respective territories. Some of these agencies were organized as subsidiaries of the French and Italian companies.

With European operations thriving, and administered from London, Murphy turned much of his attention until his 1971 retirement to expanding Rohm and Haas's presence in the rest of the world. Just as he had spent a good portion of the 1950s flying between Philadelphia and Europe, he now became a regular passenger on flights to Latin America, and to a lesser extent, the Pacific. Over the course of the decade, the existing Rohm and Haas subsidiaries in Argentina (Rohm and Haas of Argentina) and Brazil (Filibra) expanded their plants both in capacity and range of products. Murphy established new manufacturing subsidiaries in Mexico (IQASA), Colombia (Rhinco), and Venezuela (Paraplex). By the end of the decade, these five companies had combined sales in excess of $21 million. Agricultural chemicals, mainly Dithane and Stam, were even more important in Latin America than in Europe. These products accounted for forty-two percent of the five subsidiaries' combined sales. In 1969 Rohm and Haas established a Latin American Manufacturing Subsidiaries headquarters in Coral Gables, Florida, to manage and coordinate these subsidiaries. The increasing scope of Rohm and Haas operations in this region, and the success of the company's first regional headquarters in London, led Murphy to effect a decentralization similar to that which had occurred in Europe.

Elsewhere, Murphy set up manufacturing operations in South Africa (Triton Chemicals), Australia (Primal Chemicals), New Zealand (Rangi Chemicals), and the Philippines (Amfil). He established several joint ventures with native partners in countries whose laws favored this arrangement. Two such ventures appeared in India (Indofil and Modipon). A second Japanese joint venture (Jacryl) undertook the manufacture of acrylic

The company replaced its original logotype, which it shared with Darmstadt, with a new one in 1965.

products. A joint venture in Mexico (Quimica Trepic) began manufacture of ion exchange resins, as these belonged to a class of chemicals in which the Mexican government required native majority participation. Indofil involved Rohm and Haas in a secondary expansion in 1969 when it signed a contract to sell six hundred tons of locally produced Dithane Z-78 to the Soviet Union. As Canadian laws increasingly favored local manufacture, Rohm and Haas of Canada constructed a second plant at Morrisburg which produced a broader range of goods. In these ways, Murphy and his staff proved skillful at adapting Rohm and Haas's expansions to restrictions host countries placed on foreign investors.

The multitude of names under which Rohm and Haas operated around the world led to much confusion over the years. In an attempt at providing some uniform identity, the company replaced its original logotype (which it shared with Darmstadt) with a new one in 1965, a sketch of a stylized flask. In 1970 Rohm and Haas of Philadelphia purchased the worldwide rights to the name Rohm and Haas from the Röhm family and the Darmstadt company. Concurrently the Haas family sold its holdings in the German firm. The German company then changed its name to *Röhm GmBH.* On January 1, 1973, Rohm and Haas's foreign subsidiaries changed their names to match that of their parent.[15]

Despite the continued growth of foreign operations and the astonishing success of the Rhoplex emulsions, the second half of the sixties was far more difficult for Rohm and Haas than the first. More company lines—among them Plexiglas sheet, domestic agricultural chemicals, and plasticizers—suffered from stagnant or declining sales and profitability, and except for the Rhoplex emulsions, there was a dearth of high-profit, high-growth new specialties to take their places. Overall Chemicals Division sales continued to rise until the recession of 1969–1970, but at a rate no higher than the accelerating inflation. Division income plateaued at $27 million to $28 million before declining to $24 million in recession-plagued 1970. A diversification program into fibers and health products had required a significant portion of the company's construction budget and available funds, but these new investments had yet to show a profit. Although the company was solidly profitable, its prospects were not as glowing as they had appeared a few years earlier.

In the summer of 1970, in the midst of these problems and other more pressing ones surrounding the company's diversification into fibers, Haas decided it was time for a change at the top. In ten years he had moved the company ahead while maintaining its established, productive business strategies. He had developed a modern rational organizational structure, overseen the modernization of Rohm and Haas's office and research facilities, developed a new plant at Louisville, participated in a major expansion of the firm's overseas presence, and worked to diversify its product lines. For a long time he had planned on retiring early to devote his life to other pursuits. When his doctors advised him of a heart condition which would deteriorate dangerously if he continued his heavy schedule, he knew the time was right to leave.[16]

There was no Haas to succeed to the presidency. Executive Vice President John Haas, who had served his older brother as an advisor and confidant in addition to handling his staff duties in administration, did not want to run the company. He preferred to continue concentrating on people rather than products. The third generation of the family was too young and not in the business. President Haas went to his senior executives and asked each one, "If it's not to be yourself, who would you recommend as my successor, both in your own generation and the next?"[17] The name that came up most frequently was Vincent Gregory, the assistant general manager (or number two man) in Foreign Operations. Gregory was then forty-seven. At a special meeting on August 5, 1970, the board of directors approved his appointment over several other executives higher in the organization to be Rohm and Haas's new president and chief executive officer. F. Otto Haas became board chairman, and Connor edged into semi-retirement. Haas soon turned most of his energies to other activities, and he gave Gregory freedom and authority to run the company as the new president saw fit.

The selection of Gregory as Rohm and Haas's next chief

*The Baton is passed: new President Vincent Gregory and Chairman
F. Otto Haas in 1970.*

executive was surprising to some in that he was neither an
officer nor a director of the company. But Gregory was a veteran
of twenty-two years service with Rohm and Haas, and never
had worked for another firm. Most, if not all, of the people
above him in the organization were at least Haas's age, and
Haas preferred a younger man who might lead the company for
a considerable length of time.

Moreover, Gregory's career in Foreign Operations was an
asset. That was the fastest growing part of the company, claim-
ing thirty percent of total sales. Gregory had spent sixteen
years, from 1952 to 1968, in Paris and London. And as managing director of Lennig (the English subsidiary) from 1958 to
1964, and then of overall European operations from 1964 to
1968, he had run, with great independence, what was in effect

a smaller model of Rohm and Haas. He had a sense of overall perspective no one who had spent his career in the Home Office could have had. Gregory also appealed to Haas because of qualifications which he brought to the job and which Haas himself lacked. Gregory was a professionally trained manager, with a bachelor of arts degree in economics from Princeton and a master's degree in business administration from Harvard.

Gregory's elevation was the culmination of the transformation of Rohm and Haas from a family-owned and -operated business to a professionally managed corporation, a transformation which, due to Otto Haas's longevity, occurred at Rohm and Haas far later than at most comparable firms.[18] F. Otto Haas had been a transitional figure, but he remained a proprietary operator. Gregory had been chosen because of his performance with the company over more than two decades. He had no technical training but, like the founder himself, he had learned a lot along the way.

Almost immediately Gregory began instituting changes of the sort expected from a professional manager. He restructured board meetings to focus on single topics, instituted modernized, detailed, numerical analyses of possible strategies, and took steps to formalize long-range planning. Gregory proved himself to be far more inclined to make both strategic and tactical decisions personally than his predecessor had been. In that, and in his willingness to involve himself in every level of the company's operation, Gregory demonstrated a forcefulness of leadership which hearkened back to that of Otto Haas.

The company under Gregory's leadership quickly established programs to reverse the declining margins. In October 1970 Rohm and Haas announced the Efficiency Effort or Double E program, the first attempt in its history to reduce the size of its workforce. Within a year the Double E program achieved its goal of a ten percent reduction in force.[19] From a business standpoint, it was something that needed to be done. But Double E caused hard feelings as workers who always had felt secure in their jobs began worrying about losing them. Otto Haas rarely had fired anyone; he just moved people to positions where he wouldn't have to look at them, and his son had done the same.

The success of Rohm and Haas always had been dependent on a flow of scientifically advanced chemicals for industry, and Gregory was determined to continue this effort. In 1972 he announced a five-year plan to double research spending at Rohm and Haas. This was later scaled down when management concluded the increase was too rapid for profitable absorption by the Research Division. The experience taught Gregory a lesson in the complexities inherent in managing the one area over which he had had little supervisory responsibility in Europe, industrial research.

Gregory introduced tighter controls and more detailed financial analysis on proposed new construction, but then proceeded over the next several years to push construction budgets to all-time highs, as Rohm and Haas pursued a publicly announced goal to become the fastest-growing company in the chemical industry. At the April 1973 stockholders meeting, it unveiled plans to construct a massive four-hundred-million-pounds-per-year T Process plant, where propylene would be oxidized to acrylate monomers. The plant was constructed at the company's Houston facility.

For several years the company met its growth projections. Profits soared. Between 1970 and 1973, the Chemicals Division's sales rose thirty-nine percent to $394.6 million, and its net earnings rose eighty-two percent to $44.2 million. All parts of the company participated in the boom. Total corporate sales jumped to $788.6 million, and earnings rocketed by 140 percent to $65.7 million. Management planned for the future on the assumption that continued growth would make Rohm and Haas a $5-billion company by the early 1980s.

World events, however, conspired to upset these projections. In October 1973 the fourth war in twenty years broke out between Israel and its Arab neighbors. The Arabian Gulf oil producers announced a boycott of the United States in retaliation for American support of Israel. They coupled this with the transformation of a sleepy organization called OPEC (Organization of Petroleum Exporting Countries) into an effective international cartel which engineered a 370-percent increase in the price of crude oil in a few short months. American motorists began waiting in long lines for gasoline and then paying more than they ever had before. Because the synthetic chemical industry, Rohm and Haas included, used petroleum as its primary feedstock, it suffered from the same shortages and price escalations as the driver with an empty gas tank.

Rohm and Haas was squeezed between rising prices and shortages of raw materials, but it had one advantage over its competitors. The company's decision six months earlier to build the large T Process plant now appeared clairvoyant. The price of natural gas, the feedstock for the old F Process, had increased close to one thousand percent by 1976 when the pro-

pylene oxidation plant opened in Houston. Propylene was relatively inexpensive and in good supply as a by-product of
ethylene manufacture.

Rohm and Haas's total consolidated sales were up over
thirty percent in 1974, topping $1 billion for the first time, but
the entire gain was due to inflation; volume was flat. The next
year was also difficult. The nation experienced what came to
be called stagflation—a recessionary economy combined with
continued high inflation. Instead of searching for raw materials
to meet increasing demand, Rohm and Haas found itself reducing production to prevent inventory buildup. With less money
coming in to cover fixed costs, profits took a nose dive. Combined 1975 sales of the Chemicals Division and the Plastics
Division (which had been split off from Chemicals in 1974)
were down only two percent in dollars, but their earnings declined by forty-five percent.[20] There was little wrong with
chemicals and plastics beyond general economic conditions.
When conditions improved in 1976 so did sales and earnings.

At this same time, Rohm and Haas reorganized its management structure along lines recommended by a committee
formed by Gregory in 1974 and chaired by Charles Prizer, the
assistant general manager of the International Division. The
aim was to develop an organizational structure appropriate for
the $5-billion sales volume expected in the 1980s. The com-
mittee's May 1975 report recommended the company adopt a
matrix management system for all operations except fibers and
health products. Under this plan, there would be two organizational structures, one based on geography and one based on
product lines. Four regions would be created: North America,
Europe, Latin America, and the Pacific. The company's products would be grouped into four worldwide business teams:
Agricultural Chemicals; Plastics; Polymers, Resins, and Monomers (or PRM); and Industrial Chemicals. The key concept of
the matrix was that many managers would report to two
bosses, one in each structure. The regional directors would be
responsible for short-range planning, and the business directors
for medium- and long-range planning. Research remained a
corporate-wide function, but assistant research directors in
each product area would report to the business team heads as
well as to Research Division management.[21]

The company accepted the committee's recommendations
and implemented the matrix structure over the next eighteen
months. Some other major United States corporations which
adopted similar matrix structures in the 1970s later abandoned

them, complaining that dual reporting produced a paralysis of decision making. Rohm and Haas has retained the matrix, believing that it has worked well for the firm. Part of the reason for the success of the matrix at Rohm and Haas may be the strong and wide-ranging decision-making role practiced by Gregory.

By 1975 Rohm and Haas had completed the necessary but painful transformation from entrepreneurship to professional management. It had become a global operation with plants and sales around the world. Its central acrylic polymer and agricultural chemical businesses were far healthier than anyone had predicted fifteen years before. There was a new leading star in acrylic emulsions, an area which seemed well-poised for additional growth as the switch from solvent-based to water-based coatings continued. There was a large group of older but still very healthy products such as Plexiglas sheet and molding powder, Dithane fungicides, and Acryloid oil additives. The expanded research effort was beginning to yield dividends. Several products in the final stages of development—a rodenticide, an herbicide for soybean fields, an entirely new class of microbicides—gave the company cause for optimism about its future as a leading innovator in the specialty chemicals business.

But Rohm and Haas would need all of its strengths in specialty chemicals because its other major effort of the preceding fifteen years, its diversification program, was playing out to an unhappy conclusion. Speculation in 1975 and 1976 centered on the possibility that the company soon would have no alternative to taking a substantial loss and discontinuing its fibers business. Doing so, in a sense, would put it back where it was in the early 1960s as a manufacturer of specialties based mainly on acrylic polymers and agricultural chemistry. These products now would have the added burden of paying off the substantial corporate debt.

NOTES

1. *F. O. Haas interview.*
2. *John C. Haas, interview with the author, March 8, 1984, copy in Rohm and Haas Company archives.*
3. *Chandler,* Strategy and Structure, *pp. 374–78.*
4. *Rohm and Haas Company financial statements, 1960 and 1965, Rohm and Haas Company archives.*
5. *F. O. Haas interview.*
6. *Beavers interview.*

7. *Whitesell interview.*

8. *Richard Harren, conversation with the author, August 10, 1984.*

9. *Richard Harren, "Impact of Coatings Research on Rohm and Haas Business," November 10, 1980, copy in Rohm and Haas Company archives. Harren conversation.*

10. *Minutes of the Board of Directors, Rohm and Haas Company, January 26, 1968, Rohm and Haas Company archives.*

11. *Whitesell interview.*

12. *Whitesell interview.*

13. *Minutes of the Board of Directors, Rohm and Haas Company, January 23, 1968, Rohm and Haas Company archives.*

14. *Gregory interview.*

15. *Minutes of the Board of Directors, Rohm and Haas Company, June 29, 1970, Rohm and Haas Company archives. Rohm and Haas Company, "Annual Report for 1972," p. 10.*

16. *F. O. Haas interview.*

17. *Whitesell interview.*

18. *Alfred Chandler, Jr.,* The Visible Hand: The Managerial Revolution in American Business *(Cambridge, Mass.: The Belknap Press of Harvard University Press, 1977).*

19. *Gregory interview.*

20. *Rohm and Haas Company financial statements, 1974, 1975, Rohm and Haas Company archives.*

21. *C. J. Prizer, "Organization Group Study Report," May 1, 1975, copy in Rohm and Haas Company archives.*

DIVERSIFICATION

The challenges facing F. Otto Haas and Ralph Connor as they took the helm of the Rohm and Haas Company in 1960 went far beyond the need to develop a new organization. They needed, if not a new direction for the company, at least a plan for fruitfully reinvesting the firm's earnings in ways that would assure its future growth and prosperity. Over the years, the company had grown both by developing new products and by expanding existing product lines. But Haas and Connor decided that acrylic chemistry—which had fueled Rohm and Haas's expansion since the 1930s and yielded products such as Plexiglas, Acryloid, and Rhoplex—could not be expected to do so in the 1960s. What was needed, they concluded, were new products for new markets. They put their hopes on a program of diversification, one which, over time, would prove ill-advised and nearly disastrous for the company.[1]

The founder himself had been aware of the need for new markets and had begun exploring several areas in the 1950s. Among these were pharmaceuticals, fibers, building products, and films (thin layers for lamination). He seriously had considered bidding for the American branch of the Schering pharmaceutical company when the Alien Property Custodian put it up for sale in the late 1940s. He decided against it because of the financial drain caused by construction of the Houston plant and his own strong dislike of indebtedness.[2]

Fibers were another matter. Since Wallace Carothers's 1935 discovery of nylon and its subsequent introduction by Du Pont, fibers spun from synthetic polymers had become a very important part of the textile industry. Acrylonitrile, which Rohm and

Haas manufactured during World War II, had regained importance as the major monomer in Du Pont's Orlon and other acrylic fibers. Perhaps Rohm and Haas's scientists could develop a commercially viable fiber from the monomers it knew best, the acrylates and methacrylates. A small research effort begun in the 1950s showed that an interesting elastomeric (i.e., rubberlike) fiber might be spun from acrylic emulsions. It was not clear during Otto Haas's lifetime whether this fiber was technically or commercially viable, but the possibility intrigued his successors. An elastomer could give Rohm and Haas an entry into a major new area, fibers, through the same strategy, development of advanced specialty industrial chemicals, that had long been at the root of the company's success.

Connor and Haas realized Rohm and Haas knew little about making and marketing fibers. They decided the best way to acquire needed know-how in this and other areas under consideration would be to acquire small companies in the targeted industries. It was not a new idea. Companies increasingly had turned to acquisitions as a means of expanding their markets. By late 1961, Rohm and Haas had found a likely takeover candidate. It was Rhee Industries of Warren, Rhode Island, a manufacturer of both cut and spun natural rubber thread. Rohm and Haas purchased Rhee in February 1962 for $3.5 million. In Rohm and Haas's annual report that spring, Haas noted: "The competence of Rhee Industries in the production and marketing of elastomeric products will, we believe, enable us to realize additional valuable results from our research in the field of polymers."[3] Spun rubber thread such as that made by Rhee was the only commercial fiber made from emulsions, and it was from emulsions Rohm and Haas hoped to make acrylic fiber.

Problems surfaced immediately. Rohm and Haas discovered that its research fiber was nowhere near commercialization and Rhee's rubber thread faced fierce competition from the first successful synthetic elastomeric fiber, spandex. Du Pont sold one version as Lycra. U.S. Rubber marketed another as Vyrene. In this environment, natural rubber thread prices plummeted. Spandex soon captured ninety percent of the market for elastomeric thread. The sixty-seven-percent sales gain management predicted for Rhee for 1962 became a fifty-percent decline to $1.5 million; the expected profit became a $160,000 loss. These losses persisted over the following years. Rohm and Haas management learned any value Rhee might have would be in what it could teach about manufacturing fibers. As an investment, Rhee turned sour, accumulating losses which eventually exceeded the plant's worth as a development laboratory. By 1966 the company was considering closing or selling the plant. It was sold at a loss in 1968.

Rhee was just the first of a planned series of acquisitions; the 1963 capital budget had $20 million allotted for this purpose. Haas transferred Vice President Louis Klein from director of Resinous Division sales to director in charge of the Rhee Division and of long-range planning. In January 1963 Haas, Connor, and Klein selected their next target: the Sauquoit Silk Company of Scranton, Pennsylvania. Rohm and Haas completed the $4.5 million purchase on March 7, 1963, and established a subsidiary, Sauquoit Fibers Company, to hold the new assets.

Sauquoit was primarily a throwster of nylon yarn, which meant it processed the raw fibers into twisted, drawn, or textured yarn suitable for use in textiles. Haas and Connor believed Sauquoit's expertise would prove helpful in Rohm and Haas's development of proprietary fibers.[4] Sauquoit also manufactured fine-denier yarn of both main commercial nylon polymers, nylon 6 and nylon 66.

Its acquisition of Sauquoit gave Rohm and Haas access to the major fibers markets where future prospects seemed promising. Elastomers were relatively small-volume specialties, but nylon was one of the staples of the industry. Overall demand for nylon was very strong. Even a small producer like Sauquoit could (and did) make a tidy profit. Industry analysts predicted continued growth and heavy demand for nylon for many years to come.

The Rhee and Sauquoit acquisitions were part of a broader strategy of diversification. An acquisition in the films area was scuttled because of possible legal complications, and building products were dropped from consideration altogether. In health products the company made three separate acquisitions. In 1963 it acquired Warren-Teed Products of Columbus, Ohio, a small manufacturer of ethical (prescription) pharmaceuticals. In 1964 it added two manufacturers of veterinary medicines: Whitmoyer Laboratories of Myerstown, Pennsylvania, whose major products were used in the prevention and cure of poultry diseases, and a group of interlocking companies which Rohm and Haas merged into Affiliated Laboratories of East St. Louis, Illinois. Affiliated's major products were vaccines and sera for the diseases of large animals, especially for hog cholera. Despite much effort by Rohm and Haas managers and researchers, the three subsidiaries contributed only marginally to the par-

Frederick Tetzlaff, vice president for the Fibers Division

ent company's bottom line. Warren-Teed and Whitmoyer remained generally profitable, but Affiliated collapsed when the U.S. Department of Agriculture undertook a program to eradicate hog cholera.

Meanwhile, Rohm and Haas plunged into the fibers business. Sauquoit's profits and rosy predictions for the future of nylon convinced Haas, Connor, and Klein to go ahead, with or without specialties. In the spring of 1964, the company doubled Sauquoit's nylon 6 manufacturing capacity. In the fall of 1964, Louis Klein proposed spending at least $15 million on a new plant to make nylon 6. Klein said: "There is a real advantage to be gained by entering vigorously into the fiber field in order to gain the experience and the stimulation which comes from a major commitment in the general area where we are hopeful of making original contributions through research."[5] The original contribution to which Klein referred was the long-heralded acrylic elastomer, now named Fiber XFE, which was still under development in the Research Division.

After six months of deliberation and further positive forecasts, Rohm and Haas purchased the necessary nylon 6 technology and know-how from *Lurgi Gesellschaft fur Mineraloeltechnik*, a German firm which had sold similar packages to other interested parties. In October 1965 Rohm and Haas picked a plant site in Fayetteville, North Carolina, near many of the textile companies which it expected would become major customers. A Fibers Division was formed with Frederick Tetzlaff, one of the stars of the company's European expansion, as its head. The new plant would become his responsibility. Governor Dan K. Moore of North Carolina led the welcoming committee when Haas dedicated the nylon plant on June 21, 1967. The invited guests toured the modern facility. Although its primary task was to manufacture the fine deniers used in clothing, it could produce heavier deniers for carpets.[6]

President F. Otto Haas and North Carolina Governor Dan Moore cut the ribbon to open Rohm and Haas's Fayetteville plant

Unfortunately for Rohm and Haas, it was not the only chemical company to see nylon as a good investment. Among other firms entering the field were Phillips Petroleum and Dow-Badische. The latter was a joint venture of Dow Chemical and the German firm *Badische Anilin und Soda Fabrik* (BASF), one of the companies which had emerged from the postwar dismemberment of *I. G. Farben*. Not only was there increased domestic and worldwide capacity for nylon, but demand failed to keep up with industry expectations. Prices started to slip in 1966 and, by the time of the Fayetteville plant dedication, they had fallen by one-third from spring 1965 levels. As a result the new factory found it impossible to operate at a profit. In addition, process difficulties that might have been expected in a new facility kept production of first-quality yarn below expectation in the first year, further depressing the rate of return.

During the next few years, Tetzlaff drew up several plans to turn Fayetteville around. All required additional investment and all eventually failed. One idea was to switch the plant's emphasis from fine deniers to processed, textured carpet yarn. This led to a 1969 investment of $25 million to increase the heavy denier capacity to twenty-four million pounds per year (which would still leave Rohm and Haas a minor player in the industry). In addition, the company licensed a unique edge-crimping process to produce properly bulked and textured carpet yarn that could be sold directly to carpet manufacturers (instead of to throwsters). This new process improved the salability of Rohm and Haas nylon, but not to the point where it had any real advantages over the industry leader, Du Pont 501.

In seeking to develop such nylon fiber products as edge-crimped carpet yarn for marketing as specialties at higher-than-commodity profits, management was attempting to follow one of its oldest and most effective business strategies. In this case it was forcing specialties where none really existed. The Fibers Division was trying to sell the sizzle instead of the steak, and customers did not think it was worth the price.[7] The textile industry as a whole proved to be less receptive to specialties than Rohm and Haas's traditional customers had been.

By spring 1970 the company's investment in nylon exceeded $60 million, but it was losing money on every sale. The division was still supported by continued healthy profits of the older chemical and international operations. Tetzlaff, while issuing new forecasts of increased losses for the current year, predicted that 1971 finally would see fibers become profitable.

In all the excitement over nylon, Haas, Connor, Klein, and

Production of bulked and textured heavy denier carpet yarn at the Fayetteville plant

Tetzlaff had somehow lost sight of their original reason for investing in fibers. Their intent had been to use nylon as a means of learning the fibers market so they could produce and sell a proprietary fiber spun from acrylic emulsions. Instead, nylon had become a product line in and of itself and Rohm and Haas had no advantage in nylon. As a small producer it could only follow in the wake of the giants. Its inability to sell nylon was strikingly similar to its earlier failure with DDT—and for the same reason. No executive seems to have been aware of the parallel.

The development of Fiber XFE as an elastomeric yarn spun from acrylic emulsions coincided with the company's nylon misadventures. After the first emulsion returned from Rhee to the laboratory, it took until 1966 to ready a new copolymer formula for production at a pilot plant. A small facility was erected at Bridesburg to produce fiber for field testing. The pilot plant operators soon discovered the yarn they were producing lacked uniformity. More research was required. This delayed the Fibers Division's field testing. Tetzlaff remained sufficiently optimistic to apply to the Federal Trade Commission for registration of a new class of fibers, of which XFE was to be the first example. The FTC granted the petition on September 16, 1969, assigning the name anidex to the polyacrylate fibers.

Anidex was the first such listing granted since spandex a decade earlier.

XFE (its name shortened to XE) finally was ready for consumer testing. Rohm and Haas arranged for Vanity Fair, a leading lingerie manufacturer, to produce and distribute foundation garments knit from fiber blends containing XE to stores in the Philadelphia area. During December 1968 and January 1969, nine thousand of these garments were sold. Based on comments from the purchasers, Tetzlaff reported the test was "unusually successful from the point of view of the general acceptance of the fiber due to its special properties."[8]

In anticipation of a successful test, Tetzlaff had the company's engineers draft plans for an XE plant at Fayetteville. Construction of the $6.2 million facility began in January 1969. The initial installation was for a modest 750,000 pounds per year (as opposed to United States capacity of 10 million pounds per year for spandex), but there was ample room for expansion. The division scheduled plant start-up and the beginning of commercial shipments for February 1970.

It was time to introduce the fiber to the trade, but first, it needed a name. Tetzlaff selected Anim/8 because the fiber was supposed to give life to, or animate, fabrics into which it was blended. On October 22, 1969, Rohm and Haas formally introduced Anim/8 to the world in a by-invitation-only presentation at a specially constructed geodesic dome at the Tavern-on-the-Green in New York's Central Park. Amidst a quarter-million-dollar display of film, exhibits, live models, and many bolts of cloth made of fiber blends incorporating Anim/8, Rohm and Haas portrayed the new fiber as "the first really practical elastomeric fiber in textiles," because it would permit "manufacturers to add stretch to fabrics without altering their hand or appearance, and without turning them into confining rubber bands." Taking direct aim at the perceived deficiencies of the competitive spandex, Rohm and Haas listed among Anim/8's strengths its resistance to discoloration on exposure to ultraviolet light or dry cleaning solvents, and its ability to retain its elasticity after repeated washings.[9]

October 22, 1969, proved to be the high point of the life of Anim/8. Within eighteen months Anim/8's future was questionable. Within three years, the fiber had been formally abandoned. Anim/8 fell victim to at least three distinct problems, any one of which might have caused its demise.

The first was Rohm and Haas's failure to convince customers that Anim/8 was worth using. Anim/8 was twenty to thirty percent more expensive than Du Pont's Lycra spandex fiber. It took longer to recover its original length than did spandex or rubber thread. As a result, more of it was needed to impart a given amount of holding power to the fabric. The advantages which Rohm and Haas claimed for Anim/8 were not perceived by the trade as significant, especially after Du Pont introduced improved formulations of Lycra.

The second problem had to do with the changing marketplace. The largest outlet for elastomeric fibers was (and still is) women's undergarments. Anim/8 was aiming for that market, and its timing could not have been worse. It was introduced at precisely that moment when women by the thousands were throwing away their girdles forever. Elastomers never came close to meeting their sales projections of the late 1960s.[10] Anim/8's other projected market, stretch outerwear, never developed. Instead, by using stretch nylon and stretch and double-knit polyester, the industry learned to give fabrics resiliency by physical rather than chemical means, by twisting the fibers in such a way that they behaved like coiled springs.

The third problem made the first two irrelevant. Neither Rohm and Haas nor any of the textile manufacturers who tried Anim/8 were able to incorporate it successfully into commercial fabric production. Anim/8 corroded textile manufacturing equipment. No one could figure out how to stop the corrosion. Core spinning was tried. It consisted of wrapping another fiber (such as cotton or a cotton/polyester blend) around an Anim/8 filament. Core spinning failed to solve the problem because the corespun yarns consistently developed core voids, cavities in the center where the Anim/8 fiber broke.

Anim/8 had been Rohm and Haas's reason for a multi-million-dollar investment in fibers, and it had not worked. When Vincent Gregory succeeded F. Otto Haas as president in September 1970, he had to face the continuing losses in fibers. Gregory considered getting out of fibers altogether but decided against it. The Fibers Division was still delivering enthusiastic reports, and Gregory, being new on the job and full of self-confidence, wanted to succeed where his predecessors had failed. Besides, he felt he owed it to those who had put him in his new position to try and complete what had been started.[11]

Within six months Gregory approved two plans to make fibers profitable. Under the first, Rohm and Haas intensified its search for special niches within the nylon industry. In this way it hoped to avoid competing directly against the front-runners. Again, its timing was unfortunate. In the spring of 1971, the company opened a carpet yarn addition at Fayetteville and brought out a new product. It was dubbed X-Static carpet yarn, a neat double entendre because X-Static's silver-coated filaments were designed to eliminate static electricity for ecstatic homemakers. But the product came on line in a year which saw carpet yarn prices fall by twenty-five percent. X-Static's profitability suffered as a result.

Under its second plan, unveiled in April 1971, Rohm and Haas abandoned fine denier nylon entirely and converted its textile denier lines at Fayetteville to textured polyester, the "miracle fiber of the 1970s." This fiber was then in short supply. Industry analysts predicted the American market would increase from 250 million pounds per year to 600 or 700 million pounds by 1975. One outside director, banker Howard Peterson, skeptically asked the board of directors how the company expected to transform failure into success by switching from minor production of one commodity fiber to minor production of another.[12] Nonetheless, Fayetteville began producing textured polyester in July, and selling it at a profit—the first product on which the Fibers Division made money. Even with polyester the division lost $7.8 million for the full year. That was considerably more than its 1970 loss of $5.5 million.

The division continued its conversion from textile nylon to polyester and added polyester fiber texturing equipment. As the project neared completion, the profits from polyester began to outweigh the losses from nylon. The division reported its first profitable quarter ever in the fall of 1972. When the conversion was completed early in 1973, Fayetteville's polyester capacity had reached 60 million pounds a year.

With the market for polyester remaining strong, Fayetteville earned $6.5 million in 1973 on sales approaching $100 million (out of Rohm and Haas totals of, respectively, $65.7 million and $788.6 million). The profits were built on commodity-type operations. The hoped-for specialties accounted for little, but Gregory expected they would grow. Tetzlaff, approaching retirement and confident he finally had turned the division around, relinquished management of the Fibers Division to John Doyle.

Just when it looked as if the investment in fibers and the new strategies were finally paying off, the OPEC oil crisis turned everything upside down. The effect on synthetic fibers was even more devastating than it was on chemicals. In a single month, December 1973, the price of caprolactam, the petroleum-derived monomer for nylon 6, increased by fifty-five percent.

The Fibers Division found it impossible to pass on all of the increased raw material costs. Though fibers remained profitable through the first three quarters of 1974, margins suffered. Early in the year, the polyester fiber market had begun to crumble in Europe as the continent slipped into a serious recession. In August the collapse of the polyester market spread to the United States. Both price and demand for polyester yarn declined dramatically. The Rohm and Haas Fibers Division lost enough money on fibers in the last quarter to eradicate its profits from the first three. By the end of 1974 Fayetteville was operating at fifty-five percent of capacity.

The turndown in the fibers market was so rapid and so unexpected that it caught the Fibers Division, like many of its competitors, in the midst of another expansion. A second polyester unit under construction at Fayetteville would increase Rohm and Haas's polyester fiber capacity to 140 million pounds per year. The best the company could do was slow the pace of construction. Carodel Inc.—a joint venture with Teijin Ltd., a Japanese company which possessed the necessary technology—was building a polyester chip plant in Fayetteville to integrate the polyester fiber operation back one step.

When industry analysts speculated Rohm and Haas would be forced to quit the fibers business altogether and take a substantial write-off, Gregory denied it. He told the April 1975 stockholders meeting that the "probability of this situation occurring is essentially zero." Fibers Division manager Doyle said much the same thing at an October presentation to financial analysts. He stressed that the company's objective in fibers was still to move away from commodity products into specialties.[13]

Their brave talk proved to be based on an unduly optimistic forecast of an industry-wide recovery. Even as the rest of the economy and Rohm and Haas's chemical businesses began to revive in 1976, fibers remained depressed. The company lost close to $3 million on fibers in the first quarter alone, and prices kept skidding. In April it announced a retrenchment, discontinuing pioneering fibers research and curtailing the specialties development program. Polyester prices continued to fall, reaching a level in 1976 that barely exceeded the company's out-of-pocket manufacturing costs. Worldwide overcapacity exacerbated the problems triggered by recession. Doyle predicted (accurately) an $18-million divisional loss for the year. Moreover, the needs of the fiber program had led the company heavily into debt. While it no longer held to the pay-as-you-go philosophy of Otto Haas, a debt-to-equity ratio which reached eighty-one percent was not only frightening, but threatening to the company's credit rating.

On December 14, 1976, Rohm and Haas threw in the towel. Gregory announced the company was discontinuing the fibers business and was putting the division's assets, primarily the Fayetteville plant and a second one in Brazil, up for sale. It took a $40-million aftertax write-down on the value of its fibers investment. Part of the Fayetteville workforce was laid off, but the plant maintained some production of carpet nylon so that a purchaser could buy an operating plant. The write-down and operating loss in fibers, which totaled $58.8 million, exceeded earnings of the company's other divisions in 1976 by $11.8 million. Rohm and Haas reported the first losing year in its history.

It took a year to dispose of the Fayetteville plant. Rohm and Haas sold it to the Monsanto Company, a chemical firm with a longstanding program in synthetic fibers, on December 29, 1977. The seller received $55 million for a plant whose polyester installation alone would have cost $105 million to duplicate. While Gregory had hoped to get more for the plant, he was relieved just to have fibers behind him.[14] The divestiture was completed with the sale of Sauquoit to a group of the subsidiary's executives and local Scranton investors in December 1977 and, after a further write-down, sale of the Brazilian fibers and textile operations in 1978.

The failure in fibers led Gregory to reconsider the other major diversification area, health products. Although both Warren-Teed in human pharmaceuticals and Whitmoyer in veterinary products were profitable, the president concluded they could not support a research effort of sufficient size to develop new products. He decided to sell both of them. Warren-Teed went to an affiliate of Hercules Chemical and Montedison in September 1977, and Whitmoyer to Beecham Ltd. the following spring at prices that produced a net gain to Rohm and Haas of $18.5 million on the divestiture of its Health Products Division. The return from all of the divestitures went to reduce the company's indebtedness. It dropped from the unhealthy eighty-one percent debt-to-equity ratio of 1976 to a more comfortable fifty percent two years later.

With fibers behind him, Gregory remarked to an interviewer in 1979 that what top management had gotten was "an expensive education." He said he should have gotten Rohm and Haas out of fibers as soon as Anim/8 failed. The company never was able to develop the chemically unique fiber it needed for a high-margin niche in the business. Yet its past successes had been based, in large part, on finding just such niches. Failing to do so this time, it ended up pouring good money after bad.[15] But with continued strong support from the Haas brothers, Gregory was able to survive his "education," and remain at Rohm and Haas to use it to good advantage.

What Rohm and Haas management had learned was to stay within the fields and markets that it knew best. Otto Haas had kept the company profitable and growing for fifty years by adhering to a single set of business strategies and principles. He had adapted to changing circumstances and had developed new products and new markets, but he had held the firm on a steady strategic course. While one could imagine Otto Haas pushing a chemical novelty like Anim/8 (fibers research had been his idea in the first place), it is difficult to picture him manufacturing nylon or polyester, which were essentially commodity fibers. The men who directed the fibers program had thought they were applying the founder's principles, but their actions showed they did not realize the degree to which their investments were actually a major departure. Otto Haas had paid to acquire new ideas and novel products he could sell to manufacturers, but he had not tried to force his way into established fields. Successful specialties must be true scientific innovations which improve products or processes. Perhaps if Rohm and Haas's executives had consciously moved away from the founder's ideas, they might have had a chance for success. Gregory was right: it had been an expensive education.

Sticking to what it knew best was hardly the only lesson Rohm and Haas management learned during these years, nor even the most painful one. Society was changing. It was watching corporations like Rohm and Haas more closely than ever before. Of course, Rohm and Haas had been under government surveillance in two world wars. But those were matters relating to national defense and national security. Now it came under scrutiny of a different sort. The question now concerned the effect of its chemicals on the environment and on its own workers.

NOTES

Additional material for this chapter is derived from detailed minutes of the meetings of the Board of Directors of the Rohm and Haas Company.
   1. F. O. Haas interview.
   2. Connor interview.

3. *Rohm and Haas Company, "Annual Report for 1961," p. 7.*

4. *Connor interview.*

5. *Minutes of the Board of Directors, Rohm and Haas Company, October 27, 1964, Rohm and Haas Company archives.*

6. *"Rohm and Haas Opening Today,"* Fayetteville Observer, *June 21, 1967.*

7. *Whitesell interview.*

8. *Minutes of the Board of Directors, Rohm and Haas Company, January 28, 1969, Rohm and Haas Company archives.*

9. *News release, Rohm and Haas Company, October 22, 1969, copy in Rohm and Haas Company archives.*

10. Encyclopedia of Chemical Technology, *3rd ed., s.v. "fibers, elastomeric."*

11. *Gregory interview.*

12. *Minutes of the Board of Directors, Rohm and Haas Company, April 16, 1971, Rohm and Haas Company archives.*

13. *"Rohm and Haas Company Financial Analysts Meeting, October 14–15, 1975," pp. 26–29, copy in Rohm and Haas Company archives.*

14. *"Rohm and Haas Prunes Back to Chemicals,"* Business Week, *October 31, 1977, pp. 34–35.*

15. *"A Funny Thing Happened . . . ,"* Forbes, *March 17, 1979, pp. 64–65.*

A NEW CLASS OF CONCERNS

Over the course of its first half century, Rohm and Haas had developed a set of relationships with the government and with society at large that engendered a fair amount of company pride. Otto Haas operated his company as a good corporate citizen and it obeyed the laws governing corporate behavior. It took great interest in the well-being of its employees and felt a strong responsibility toward them. It paid good wages, offered solid job security, and provided such amenities as company clubhouses for after-work socializing. For all of these reasons, Rohm and Haas was known as a good place to work.

This was true even though the work in the Rohm and Haas plants was sometimes unpleasant. Industrial-scale chemical processes often gave off oppressive heat and odors. Workers could be exposed to chemicals which irritated the skin or other parts of the body, but in those days it was the general belief of workers and management that putting up with such conditions was just part of the job. Science (and by inference science-based industry) generally was perceived as a progressive force of un-alloyed good. In any case, at worst, the workplace conditions at Rohm and Haas were considered uncomfortable at times.[1] No one seemed to consider the prospect that these conditions might conceal a more serious, long-term threat to worker health. Questions were never raised about that, nor about waste disposal, air quality, or the host of other concerns now grouped under the rubric of the environment. It was not that Rohm and Haas turned its back on these issues, but that no group in industry, government, or the scientific community was confronting them.

At Rohm and Haas, as at many other industries, there was always danger of industrial accidents, fires, explosions, and other hazardous situations. Rohm and Haas tried to keep these potential problems under control. Its plants had their own fire brigades and rescue squads. Goggles and other safety equipment were available to anyone who asked for them. In those few instances where a process used a known toxic chemical—the most striking example being nickel carbonyl, an acute poison used in the Houston F Process plant—company engineers designed closed processes to prevent worker exposure. With nickel carbonyl, Rohm and Haas even sponsored development of an antidote (by Dr. William Sunderman of Jefferson Medical College), which the company kept in stock and supplied to other users of nickel carbonyl as needed.[2]

New chemicals were coming into use every year, however, and little was known of their toxicity. As a rule, Rohm and Haas, like the chemical industry in general, simply did not investigate the long-term toxicity of its products. One expert in public health felt compelled to explain this to a congressional committee as late as 1976: "This lack of recognition of occupational disease and the effect of the microchemical environment also stems from the absence of necessary studies by government, by industry and by the scientific community."[3] Chemicals were widely considered benign (or at least harmless) unless shown to be acutely toxic.

One typical chemical process at Rohm and Haas's Bridesburg plant was the manufacture of ion exchange resins. Ion exchange resins are polymers which have the ability to replace an undesirable ion (or charged particle) in a solution with one which is unobjectionable. Household water softeners, where hardness-producing calcium ions are replaced by soluble sodium ions, are a well-known application of this process. The original Amberlite synthetic ion exchange resins had been introduced by the old Resinous Products Company affiliate in 1940. The resins remained small-volume exploratory products until the early 1950s when a new generation of resins, first announced by Rohm and Haas in 1948, began to achieve significant sales. This new line of vastly improved resins was the result of intensive Rohm and Haas research by a group led by Dr. Robert Kunin. Among this new generation was the world's first strong-base exchange resin, Amberlite IRA-400. IRA-400, when used in conjunction with a strong-acid exchange resin, made possible the production of extreme high-purity deionized water, needed for many industrial uses, including high-pressure power plants and pharmaceutical manufacture.[4]

IRA-400 had been made in the laboratory by reacting polystyrene polymer beads successively with bis-chloromethyl ether (BCME) and a tertiary amine. BCME was made in the laboratory from formaldehyde and hydrochloric acid. After the company's 1948 announcement, production moved to the semi-works. This was a halfway house between research and production. Here a new product or process could be scaled-up to commercial quantities, and sufficient amounts could be made for test-marketing. In the semi-works, the company experimented with using chloromethyl ether (CME), which contained one instead of two chlorine atoms per molecule, as an alternative to BCME. By 1950 the company concluded CME was the more efficient chloromethylation agent. Thereafter, the semi-works produced only technical grade CME. The CME thus produced was far from chemically pure. It contained a small amount of BCME as a by-product. As the semi-works engineers solved the problems of scale, and the sales of the new ion exchange resins increased, chloromethylation was shifted to Building R-11 and, in 1954, chloromethylation and the production of CME moved to a new dedicated line in Bridesburg plant Building 6. Building 6 was a sprawling shedlike structure which the CME kettles would share for over a decade with several other processes.

A pair of kettles in Building 6 produced CME. One generated hydrogen chloride gas by addition of chlorosulfonic acid to water. Several steps were required to produce CME in the second kettle. First liquid methanol was added, then solid paraformaldehyde and a solid catalyst were shoveled in, and finally the kettle was closed to receive the hydrogen chloride gas from the first kettle. These chemicals reacted to form CME, which was then decanted to a storage tank. Other kettles served to react CME with copolymer beads. The resulting batch contained chloromethylated beads and excess CME. Water was added on top of this mixture to destroy the excess CME. The CME-water reaction was violent and on occasion put chemical fumes into the building atmosphere. The beads were then sent to another building, where they reacted with trimethylamine to form the finished ion exchange resin. The Building 6 workers inhaled fumes from the frequently opened kettles. The fumes irritated the lungs and corroded clothes, but the long-term health perils were not recognized. Workers themselves later recalled having been little concerned.[5]

In the fall of 1962, Rohm and Haas management had its first inkling that something might be wrong in Building 6. Accounts differ on whether it was the area supervisor, the plant personnel director, or the plant physician who first became suspicious. However, a report reached the Washington Square Home Office in November saying there had been three recent cases of lung cancer among Building 6 and semi-works employees. Acting on this report, the company invited Dr. Katharine Boucot Sturgis, a renowned lung disease specialist and professor of preventive medicine at Woman's Medical College of Pennsylvania, to discuss the problem before a group of plant and Home Office officials and recommend a course of action. At a meeting on November 27, she offered to add one hundred Bridesburg employees, both from Building 6 and (to avoid hysteria) other plant areas, to her ongoing research program: the Pulmonary Neoplasm Research Project. This long-term project included chest x-ray examinations of Philadelphia area workers. On the 29th, John Mitchell, the assistant plant manager, read a letter to the Building 6 employees describing his and the company's "concern over the cases of cancer that have appeared among Building 6 personnel" and introducing the new x-ray program. He also expressed his opinion that a connection between Building 6 chemicals and cancer was a long shot.[6]

The company had no idea at the time, which, if any, of the many chemicals used in Building 6 or elsewhere might be a suspect carcinogen, or if the deaths were merely statistical happenstance. The whole idea of environmental carcinogenesis was new. It was not well-understood by the chemists at Rohm and Haas, or by any but a few outside specialists. Rohm and Haas, like most chemical companies of the era, did not have a toxicologist on its staff. So, at management's direction, the plant prepared a list of 102 chemicals. This list was sent for comment to the Sloan-Kettering Institute in New York, perhaps the nation's leading cancer study center. Sloan-Kettering replied that it recognized no known or suspected carcinogens among the 102 chemicals. Sending the list to Sloan-Kettering was later criticized by an independent cancer expert on the grounds that the famed center's expertise was in cancer treatment and cure, not causation.[7] This criticism, if valid, chiefly demonstrates the extent to which Rohm and Haas management was moving into an area foreign to its collective skills.

The third action the company undertook was to intensify the engineering department's development of closed processes, thereby limiting worker exposure in Building 6 and elsewhere in the plant.[8]

The production lines kept operating as the various investigations and process changes proceeded, and, as might be expected with a disease whose incubation period was measured in decades, additional lung cancer deaths occurred. In July 1964, Sturgis, who had become more concerned by the continuing fatalities, recommended Rohm and Haas try to shorten the list of suspect chemicals and sponsor animal carcinogenicity studies on those remaining. She further recommended that Rohm and Haas approach Dr. Norton Nelson, director of the New York University Institute of Environmental Medicine. Sturgis advised that Nelson's institute was the only organization properly equipped to conduct the animal carcinogenicity tests she thought necessary, so novel was the use of animals to test for carcinogenicity in the mid-sixties.

Nelson visited Bridesburg that month and expressed his interest in undertaking such a study. In August, Dr. Benjamin Van Duuren, NYU's expert in industrial toxicology and a former Du Pont researcher with a doctorate in organic chemistry, followed his chief down to Bridesburg. There, from a long list of chemicals used in the plant, Van Duuren selected approximately twelve chemicals, including BCME and CME, as likely suspects. After some discussion with Rohm and Haas officials, including Chairman and Vice President for Research Connor, on the proper scope of the work, Nelson, on January 21, 1965, submitted to Rohm and Haas a protocol for the study he would undertake of these dozen suspects. Rohm and Haas agreed the proposal was sound, and on March 1, asked Nelson to draw up a formal agreement.

On April 1, Nelson traveled to Pennsylvania for what he believed would be a routine contract signing, but there was one hitch. Nelson's draft agreement gave him the sole and exclusive right to decide what would be published, and where. Rohm and Haas officials objected to this clause in the agreement. Nelson responded that without full publication rights he would not do the study. The matter ended there. Perhaps Nelson and Rohm and Haas had come up against the mutual mistrust of academics and businessmen resulting from differences between one culture which values the unfettered exchange of ideas, and another which prizes proprietary information. A more cynical observer might see in this disagreement an attempt by a corporation to hire and control outside expertise. But Dr. Sturgis later commented: "It led me to believe Norton Nelson didn't understand the community-mindedness of Rohm and Haas."[9]

Rohm and Haas next approached Hazleton Labs of Falls Church, Virginia, a well-respected independent laboratory with extensive experience in animal toxicity studies. By the end of July 1965, Hazleton had drafted a protocol for the research, Rohm and Haas had accepted it, and the two companies had signed a contract for the work. Hazleton was one of the best facilities of its kind, but neither it nor any other institution had particular expertise in chemical carcinogenesis equal to the NYU group. Thus, while Van Duuren had intuitively selected likely culprits, it took Hazleton until December 1966 to develop its own list of twelve suspects from the original roster of 102 chemicals. Using a strain of tumor-prone mice, Hazleton then began a short-term laboratory test on ten suspect chemicals. This test was designed to finger a probable culprit, rather than produce quantitative data, and it accomplished its goal. In early February 1967, Dr. Hazelton reported to Rohm and Haas the one chemical which showed suspicious results was BCME. Exposure to BCME produced larger numbers of lung adenomas (benign tumors without a direct human counterpart) in the test animals than in the control group.

The chemists who ran Rohm and Haas seemed unsure of the appropriate conclusion to draw from this study. Did the fact that BCME caused benign tumors in tumor-prone test animals prove that it would cause lung cancer in workers? The connection was not clear to them. Their uncertainty reflected a disagreement which had persisted among scientists for at least a decade, and continues to this day: Of what value are animal tests, as opposed to human epidemiological studies? In part, the Rohm and Haas executives were led by their own training and experience as chemists and chemical industrialists to continue clinging to a long-held implicit assumption that chemicals could be assumed innocuous until proven otherwise. They also suffered from the absence from the Rohm and Haas staff of any scientists with the appropriate backgrounds for interpreting the tests.

While these negotiations and experiments were proceeding, the company was changing the CME/anion exchange resin process to reduce worker exposure. In March 1962, before the first word of Building 6 cancers reached management, company engineers had begun development work to replace the solid catalyst with a liquid one that could be fed into the kettle through a feed pipe, eliminating a major reason for opening the kettle and exposing the workers to fumes. In March 1964 they installed a distillation system to recover and recycle some of the unreacted CME from the chloromethylated beads. With less CME in the kettle, the addition of water during the quench process was a quieter reaction, thereby reducing the potential for the escape of fumes. In July 1965 the company moved chloromethylation of the beads to a new, larger kettle in the first bay of a new section of the building, known as 6B. Eventually, 6B became a separate building with concrete, isolated floors and improved ventilators to handle any vapors which might escape through leaks in the system. In April 1966 a reverse quench tank was installed in 6B, allowing the chloromethylated beads to be dumped, within a closed system, into a large tank of water to destroy unreacted CME without fumes escaping into the building. Rohm and Haas engineers then replaced the solid paraformaldehyde with methyl formcel, a liquid which could be fed into the kettles without opening the manholes (1968).

While these changes were occurring, company researchers were working on an entirely new process for the CME used in the production of ion exchange resins, one in which the CME would be produced continuously, rather than in batches. The engineering was completely different, but the underlying chemistry remained the same. Pilot plant studies of the new process occupied most of 1969, and a full-scale continuous production facility followed in 1970.[10]

Meanwhile, Nelson, Van Duuren, and their colleagues at NYU had not forgotten about the aborted Rohm and Haas protocol. As Van Duuren continued to ponder the list of suspects he had composed the previous year, his suspicions became increasingly focused on two chemicals, CME and BCME. When the National Cancer Institute invited him to propose a research project on any topic he chose in the spring of 1966, he chose CME and BCME. Within nine months of receiving the institute grant in July 1966, he had produced a high percentage of skin tumors in mice with BCME. In September 1967, he presented a short preliminary report of his work to Nelson, who arranged for its publication. Nelson, knowing the primary route for plant worker exposure to BCME was through breathing, put a second group of NYU scientists to work on inhalation studies in rats. This study went more slowly, but by May 1971 it had made a major discovery: as little as one-tenth of a part per million (0.1 ppm) of BCME in the air, at an exposure rate of just six hours a day, produced lung cancer in rats. BCME was the most potent carcinogen ever tested at the NYU labs. Nelson called Rohm and Haas to tell the company what his researchers had found, and to invite Rohm and Haas to send someone up to Manhattan to examine the institute's results. A team, consisting of Associate Director of Research H. Fred Wilson, the company's top agricultural chemist and a man therefore familiar with biologically active substances, and Jack N. Moss, Rohm and Haas's senior pharmacologist, went to NYU on June 1.

Wilson reported back to the Rohm and Haas management that the NYU work was striking. Not only had the test animals developed respiratory cancers similar to human lung cancers from these minuscule doses, but the cancers had metastasized to other organs in many of the rats. No one at Rohm and Haas had had any reason to suspect that BCME (or any other chemical) might be carcinogenic at such a low level. Vincent Gregory, who had been president of the company for less than a year, recalled that Nelson's report was a great shock: "I sat with my research people, and then it hit me: that [0.1 ppm] was a teaspoonful in a whole city block."[11]

At a special Operations Committee meeting on July 1, Gregory, in an unprecedented move, closed the CME plant for retrofitting. His aim was to reduce possible exposures to well below the 0.1 ppm dose level that had caused animal cancers. He instructed the Research Division to develop a monitoring system for BCME of a sensitivity of at least one part per billion (0.001 ppm). At a meeting at his home on July 5—a holiday— he helped Assistant Bridesburg Plant Manager Edward Riener prepare a statement outlining the problem to the workforce. In a series of meetings on July 7, Riener informed some eight hundred men who might have been exposed to BCME in the plant that "the seriousness of the NYU findings has prompted the company to assume that exposure of people to CME [which contains BCME] is undesirable."[12]

It took six weeks to install the necessary modifications in the CME plant. These changes actually were minor compared with those emplaced over the previous decade; the new expenditure was $175,000, versus a cumulative total of $6.5 million for the earlier work. The 1971 modifications consisted mainly of airmask supplies, additional partitions, more ventilation equipment, and remote controls.[13] The company arranged for exposed workers to be regularly tested for evidence of lung cancer by specialists at Johns Hopkins University in Baltimore. The changes themselves may have been minor, but they were undertaken with a new urgency. Gregory's approach was to keep the plant closed until all the needed changes were made.

The biggest question remaining to be answered was what significance the NYU tests had for human carcinogenicity. Here Rohm and Haas's top scientists—chemists like Vice President for Research Ellington Beavers and Wilson—found themselves in the middle of the long-running controversy over the value and relevance of animal experiments in predicting human carcinogenicity. The NYU work was far more extensive and persuasive than the Hazleton research had been, but it was still animal work. Gregory ordered an extensive epidemiological study of the Bridesburg workforce to determine what relationship there was between BCME exposure and human lung cancer. Several years later, Beavers was asked at what point he became convinced BCME had been responsible for the lung cancer deaths. He replied "1974," when the epidemiology study was completed.[14] On a practical level, though, the date when the Rohm and Haas management accepted this conclusion was of little import since it had taken steps previously to combat the carcinogenicity of BCME.

The protective measures the company took between 1962 and 1971 were of little use to the workers who had had prior exposure, however, and with the long incubation times typical of cancer, exposed workers and retirees continued to die of the disease. No one knows for certain how many Rohm and Haas employees have died from BCME-induced cancers, but by 1976, Rohm and Haas acknowledged twenty-seven deaths as BCME-related (by offering workmen's compensation settlements), and by 1984 the number had risen to fifty-one.

Because men continued to die and scientists continued to investigate BCME, the story did not end with the 1971 actions, or even the 1974 study. On October 2, 1974, the Health Research Group, a public interest organization founded by consumer advocate Ralph Nader, held a press conference in Philadelphia to publicize findings of its own. It charged that "the corporate management of the Rohm and Haas Company in Philadelphia suppressed for years its knowledge of human lung cancer at its Philadelphia [i.e., Bridesburg] plant where CMME [sic] was produced. Because of this suppression, workers at the plant did not know until recently that they were exposed to a carcinogen."[15] The group further charged that the company recognized BCME to be a human carcinogen in 1967, but failed to take adequate protective measures until required by law in 1973. These allegations were widely reported and followed in the national media. On February 26, 1975, CBS devoted a fifteen-minute segment of its nationally televised news program "CBS Magazine" to Rohm and Haas and BCME. On October 26, 1975, a long, sensationalistic account appeared in *Today*, the magazine section of the Sunday *Philadelphia Inquirer*, under the

This article in the Philadelphia Inquirer *Sunday supplement reported on the health problems at the Bridesburg plant.*

title, "54 Who Died." Two years later, the authors of this article produced a book, *Building 6*, out of the same research.

What had been an internal Rohm and Haas tragedy now became the subject of a loud chorus of public criticism. Rohm and Haas, whose management always had prided itself on a paternalistic concern for its workers, was portrayed as a cold, uncaring corporate exploiter. This public crisis was one for which Rohm and Haas was poorly equipped and ill-prepared. Management had no experience in talking to the press. The company did not even have a public relations department; what little it did in the public arena was handled as a sideline by its Advertising Department. Gregory discovered that it was no longer possible for a chemical company to operate with the kind of public anonymity that Rohm and Haas long had enjoyed. Americans of the 1960s and 1970s, reared with the crises of Vietnam and environmental problems appearing nightly on their television screens, were vocal and concerned. In 1976 Rohm and Haas created a Public Relations Department to deal with a whole host of concerns which had not existed a few years before.

The tragedy taught Rohm and Haas some other lessons as well. It learned to assume exposure to chemicals in the workplace might well be harmful, and plants should be designed to prevent exposure. Chemicals in the workplace were to be considered guilty until proven innocent. The company also discovered the importance of gaining expertise on the toxicity of its chemicals. In 1970 it employed no toxicologists or epidemiologists and few corporate health and safety professionals. Fifteen years later, it had more than a hundred. Meanwhile, the aftereffects of the BCME episode persisted. Over a dozen suits filed by Bridesburg employees and their families were still pending against the company in 1984.

Rohm and Haas was one of many chemical manufacturers that experienced worker-health problems in this period. Other chemicals—benzene, vinyl chloride, asbestos—caused wrenching tragedies elsewhere. As concerns mounted, federal legislation was enacted to stiffen controls. These laws included the Occupational Safety and Health Act of 1970 and the Toxic Substances Control Act of 1976.[16] Gregory, having learned his lesson from BCME, was one of very few corporate executives to testify in Congress in favor of the latter act.

As ecological and worker-health concerns increased, agricultural chemicals came under intense scrutiny. Questions were raised about the possible harmful side effects of insecticides, herbicides and fungicides. A new federal regulatory agency, the Environmental Protection Agency (EPA), appeared in 1970. Two years later, the Federal Environmental Pesticide Control Act of 1972 (FEPCA) gave the EPA sweeping authority to control agricultural chemicals.

Public concern about pesticides was stirred by Rachel Carson's *Silent Spring*, a controversial, 1962 best seller on the perils of DDT. Once she sounded the warning that DDT was harming non-target insects and birds, others took up the cry. The result was a ten-year battle which ended in 1972 with a government ban on DDT (and its little-used Rohm and Haas cousin Rhothane). By that time, it was well-understood all pesticides would require careful scrutiny. Additional pesticides followed DDT off the American markets, and new use restrictions

Industrial hygiene demonstration for a group of Rohm and Haas employees

were placed on others. Rohm and Haas was only minimally affected at first because it had stopped making DDT in the mid-fifties. But FEPCA necessitated the reevaluation of its established agricultural lines.

One of the products subjected to such scrutiny was Tok, a selective herbicide which the company introduced in 1962. In four years, Tok established itself as the only herbicide which could replace hand-weeding in fields of cole crops such as broccoli, cauliflower, and brussels sprouts. Tok was not a major product as agricultural chemicals go. Domestic sales in the 1970s fluctuated between 1.5 and 2.1 million pounds per year, about one-tenth as much as Dithane. But Tok was important in its special niche. Although most of its production was used on crops in California, Tok was sold elsewhere in the United States and overseas, especially in France where it was effective against weeds in wheat fields.

As part of a general study of hundreds of chemicals, the National Cancer Institute (NCI) undertook a study of the carcinogenicity of nitrofen (the active ingredient in Tok) in 1975. In 1977 it reported preliminary results suggesting nitrofen was

both carcinogenic in high doses in laboratory mice and terato-
genic (birth-defect producing) in female mice. The NCI issued
its full report the following spring. Before then, however, Rohm
and Haas had brought the matter to the EPA's attention. The
company undertook further tests of its own, and issued recommendations to Tok distributors and users that women of
childbearing age not be exposed.[17] Rohm and Haas argued the
carcinogenicity tests were not very convincing because of the
high levels of nitrofen needed to demonstrate carcinogenic effects, and because of the years of use without reported illness
in plant or agricultural workers. The company's arguments
were found insufficient rebuttal by several countries, including
Sweden and Norway. They began removing Tok from their markets after the NCI study was published in March 1978. The
concern was centered on the health of agricultural workers;
Tok rapidly degrades, leaving no detectable residues on the
produce or in the soil.

Rohm and Haas began still other tests on Tok. By early 1979
its research effort discovered two troubling facts which the
company duly reported to the EPA. It found Tok was absorbed
readily through the skin, thus exposing workers to far more
significant doses than previously had been believed. It also
found exposure to Tok could produce birth defects in experimental animals. Rohm and Haas asked the EPA to restrict
Tok's use to certified applicators wearing protective clothing,
and to ban its use by women of childbearing years.[18]

Initially, these new restrictions seemed acceptable to both
the EPA and the California State Department of Food and Agriculture. In April 1980 California restricted the use of Tok in
line with Rohm and Haas's recommendations of the previous
year. But the decision, and particularly its prohibition against
women applying the pesticide, drew objections from many Cal-
ifornians, ranging from farm labor organizations to the chief of
the Pesticide Division of the California Occupational Safety
and Health Administration. The restricted use of Tok challenged an important societal goal, that of providing equal employment opportunity for women. Critics also noted that Tok's
safety to men was at best unknown. Some thought it should be
totally banned.

The California Department of Food and Agriculture spent
most of the summer of 1980 weighing the pros and cons of Tok.
Rohm and Haas argued that the pesticide's economic benefits
made the male-only policy a reasonable one. By August it became evident the sex-discrimination argument would carry the

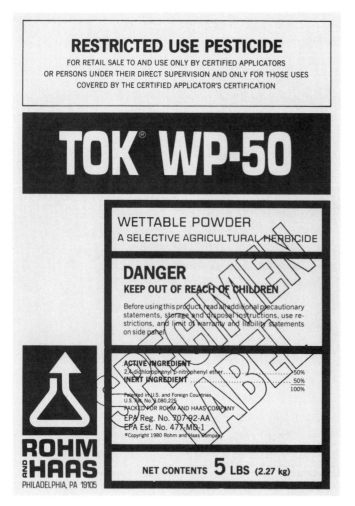

Rohm and Haas briefly used this label on packages of Tok, warning that it was unsafe for use by women of childbearing years.

day. Rohm and Haas voluntarily withdrew Tok from the American market pending development of an application method which would provide equal protection to both sexes.[19] By this voluntary action, the company hoped to be able to reintroduce Tok and protect its sales in those countries whose governments would accept the restrictions for female applicators.

Under pressure from the EPA, Rohm and Haas abandoned its attempt to reintroduce Tok domestically in September 1983, and requested that the EPA cancel the herbicide's registration.[20] As a result, Tok continues to be manufactured and marketed solely by Rohm and Haas France, where its use in heavily mechanized wheat farming (compared to the hand-labor involved with cole crops) by male applicators only was approved by the French authorities.

In recent years, virtually every agricultural chemical has come under similar scrutiny. Fortunately for Rohm and Haas, not all have suffered the same fate. On August 10, 1977, the EPA issued a Rebuttable Presumption Against Reregistration (RPAR) against ethylene bis-dithiocarbamate (EBDC) fungicides, the leading example of which was Dithane, the long-time mainstay of Rohm and Haas agricultural chemistry. The EPA issues an RPAR when it uncovers evidence suggesting a substance might be harmful to health or to the environment. The procedure allows proponents of the challenged substance to submit both toxicological and economic benefit arguments to rebut the concerns. Existing uses of Dithane were permitted to continue while the RPAR process was underway, but no new ones could be introduced.[21] After five years of scientific study, economic analysis, and discussion with the interested parties, the agency issued a final determination. It said EBDC use could continue subject to minor labeling changes to protect applicators and the agreement by Rohm and Haas and the other rebutters to undertake additional scientific studies.[22] Even a final document such as the EPA's 1982 statement on EBDC may not conclusively resolve the issue. In 1984, under the pressure of lawsuits filed by the AFL-CIO and the National Resources Defense Council, the EPA agreed to reopen its investigations of EBDC and several other pesticides.[23]

In the cases of CME, Tok, and Dithane, the concerns raised were largely ones affecting particular groups of workers. By contrast, hazards resulting from air and water pollution and waste disposal could place the general population at risk. Rohm and Haas has devoted increasing attention in recent years to preventing pollution, developing alternatives to traditional landfill disposal for potentially hazardous material, and assuring the communities in which it does business that it takes proper precautions.

Indeed, for Rohm and Haas and for the chemical industry in general, such concerns have become a permanent component of business operations. They are never-ending concerns, for as the state of scientific knowledge changes, the definitions of safety, harmlessness, and acceptable risk change as well. Prevention of exposure to acute toxins was good practice in the 1940s; today we know more and worry more about carcinogenicity and other chronic risks. Synthetic pesticides were once considered the ultimate solution to crop protection; today we know they can create problems as well. As scientific under-standing evolves or as societal demands and government poli-cies change, chemicals once considered safe may be banned or reclassified as hazardous. Landfills were once considered the ideal way to dispose of hazardous wastes safely; today we know that chemicals placed in landfills can leach into neighboring areas over a period of time. Operating a chemical business will never again be free of these concerns, or be as simple as it had been in the past.

For Rohm and Haas, events and changes of the 1960s and 1970s gave a new dimension to Otto Haas's imperative that his company know the chemistry of its products and the nature of its markets better than anyone else. The hard and tragic lesson the company learned from the lives lost in Building 6 was that the moral imperative of researching how its chemicals could affect workers, the public, and the environment is as important as the economic imperative of researching how the chemicals improve the customers' products.

NOTES

Additional material on the history of CME comes from Willard Ran-dall and Stephen Solomon, Building 6: The Tragedy at Bridesburg *(Boston: Little, Brown and Company, 1977) and Rohm and Haas Company, "Rohm and Haas Company Comments on Some of the Major Errors and Distortions in 'Today' Magazine Article—October 26, 1975—'54 Who Died,' " n.d., Rohm and Haas Company archives.*

1. Randall and Solomon, Building 6, *pp. 59–60.*

2. Michael Seidel, conversation with the author, September 12, 1984.

3. U.S. Congress, House Manpower and Housing Subcommittee of the Committee on Government Operations, Hearings on Control of Toxic Substances in the Workplace. *94th Congress, 2d sess. State-ment of Dr. Thomas Manusco, Research Professor of Public Health, University of Pittsburgh, May 11, 1976.*

4. Robert Kunin, Elements of Ion Exchange *(New York: Reinhold Publishing Corporation, 1960).*

5. At least according to Bridesburg Building 6 plant worker Bob Pontius, as cited in Randall and Solomon, Building 6, *pp. 58–59.*

6. Paul Grubb to Lloyd Covert, November 29, 1962 and attached memorandum, J. C. Mitchell, "Discussion of Chest X-Ray Program with Plant Personnel," November 29, 1962, Rohm and Haas Company archives.

7. Dr. Norton Nelson, New York University Institute of Environ-mental Medicine, quoted in Randall and Solomon, Building 6, *p. 75.*

8. L. E. Westkaemper, "The Development of Process Safeguards at the Philadelphia Plant in the Period 1962–1971," Rohm and Haas Company archives.

9. Randall and Solomon, Building 6, *pp. 98–99, 101.*

10. Westkaemper, "Process Safeguards."

11. Randall and Solomon, Building 6, *p. 126.*

12. *"Information read by Dr. Riener to Philadelphia Plant Employees—July 7, 1971," Rohm and Haas Company archives.*

13. *Westkaemper, "Process Safeguards."*

14. *Televised interview, "CBS Magazine," CBS Television Network, February 26, 1975.*

15. *Andrea Hricko and Daniel Pertschuck,* Cancer in the Workplace: A Report on Corporate Secrecy at the Rohm and Haas Company, Philadelphia, Pennsylvania. *(Washington: Health Research Group, 1974), p. 2.*

16. *Samuel S. Epstein,* The Politics of Cancer *(San Francisco: Sierra Club Books, 1978), pp. 79–149.*

17. *"Tok," Memorandum and Attachments, Patrick J. McNulty, September 16, 1977, Rohm and Haas Company archives.*

18. *"Tok Regulatory Action," Memorandum, R. E. Petruschke, June 19, 1979, Rohm and Haas Company archives. S. F. Krzeminski to Robert Taylor, EPA, June 15, 1979, Rohm and Haas Company archives.*

19. *R. E. Petruschke to Edwin Johnson, EPA, August 21, 1980, Rohm and Haas Company archives. "Pesticide Linked to Birth Defects,"* San Jose Mercury-News, *September 28, 1980.*

20. *Patrick McNulty to Edwin Johnson, Director, Office of Pesticides Programs, EPA, September 15, 1983, Rohm and Haas Company archives.*

21. *42* Federal Register *40618.*

22. *47* Federal Register *47669.*

23. *National Resources Defense Council Inc., et. al. vs. United States Environmental Protection Agency, Civil No. 83-1509, D. C. Settlement entered September 20, 1984. "EPA Agrees to Reconsider Production of 13 Pesticides,"* Philadelphia Inquirer, *September 21, 1984.*

CHAPTER 13

REDEDICATION

An immediate problem facing Gregory and his management team in 1977 was restoration of the company's financial health in the wake of accumulated losses from fibers, and slippage elsewhere caused by inflation, raw material cost increases, and a company strategy that had been based on expectations of high growth. Gregory recognized changed economic conditions made a fundamental shift in corporate objectives necessary. At the company's 1978 annual meeting, he announced adoption of a new central standard for measuring progress, one that emphasized earnings rather than revenues. The new standard came to be known by its acronym, RONA, for Return On Net Assets. Gone was the prediction of $5 billion of sales for the 1980s; in its place was a prediction of a ten-percent RONA (up from seven percent). Management expected this improvement to double the per-share earnings of Rohm and Haas stock. Profitability replaced growth as the company watchword.

In the first of a series of management changes announced at the stockholders' meeting, John Haas stepped down as chairman of the board four years after succeeding his brother in the post. Gregory was named board chairman while continuing as chief executive officer. Donald Felley became president and chief operating officer, the latter a new title at Rohm and Haas. Felley acquired primary responsibility for the firm's day-to-day operations.[1]

Like many other Rohm and Haas executives, Felley had been trained as a chemist, receiving a Ph.D. from the University of Illinois in 1949. Like Gregory, he had spent much of his career overseas, serving as manager of Minoc, Rohm and Haas's

Donald Felley, president and chief operating officer

French subsidiary, from 1958 to 1964. He had succeeded Don Murphy as head of the International Division in 1971. After the matrix reorganization of the mid-1970s, he had become regional director for North America.

With its new emphasis on profitability, Rohm and Haas placed a tight rein on staffing and refined strategic planning to improve its assessment of the risks involved in proposed projects and also the potential payoffs. While seeking to contain costs, however, it increased its spending for new product research in an attempt to stay ahead of its competition. Significant research efforts also were dedicated to improving current processes and developing new ones for the manufacture of established products.

Successful introduction of new chemical specialties was an integral part of the RONA-based strategy. They tended to have profit margins higher than established products. That was because they could be sold on the time-honored Rohm and Haas basis of making a unique improvement in customers' products, rather than on the basis of costs. This reinvigoration of the Research Division, increased emphasis on coordinating research with marketing, and attention to chemical specialties began to pay off.

Vacor rodenticide was one such product for which Gregory had high hopes. When this rat poison was introduced in 1975, he predicted it would take over the rodenticide market and contribute $13 million in annual net profits by the early 1980s. Vacor was Rohm and Haas's first major research-generated product line of the 1970s. It came out at a time when mounting fiber losses and new product failures made management particularly anxious for a success. Vacor killed rats with a single dose, whereas the dominant rodenticide, warfarin, required a week of daily ingestion. Although Vacor was originally a product of the company's animal health research and had been marketed by the Whitmoyer Laboratories subsidiary, it was withheld from the 1977 divestiture and was reassigned to the Agricultural Chemicals Business Team. It never lived up to expectations, and the company withdrew Vacor in 1979, accepting a $15-million loss.

The problem with Vacor was it proved to be toxic to humans. Its human toxicity was unknown prior to its commercial introduction in South Korea in 1975. Extensive tests on animals had shown Vacor to be relatively safe for non-target species including rhesus monkeys and baboons (two species chosen as good stand-ins for humans). After it was introduced in South Korea, authorities there, along with Rohm and Haas scientists, traced several fatalities to ingestion of Vacor. The product had gained United States registration from the Environmental Protection Agency in 1975, but it was continually plagued by questions concerning its human toxicity. Additional problems resulted from the decision to market Vacor directly to consumers as a finished product (in small packets ready to be placed where they could be found by rats). This was a type of marketing with which the firm was unfamiliar. When Vacor's continued use was challenged by California state regulatory authorities in 1979, the company decided it should no longer be defended.[2]

Other innovations fared much better. From its earliest research days, Rohm and Haas had been looking for a chemical to control a major American cash crop. The absence of such a mainstay had long put a crimp in its agricultural chemical efforts. Dithane, its top agricultural product, was used more widely overseas than in the North American market.

By 1975 the company appeared to have found its long-sought agricultural champion in an experimental chemical known as

RH-6201. RH-6201 showed tremendous promise as a post-emergence (after the plants appeared) herbicide for the treatment of broadleaf weeds in soybeans, the nation's second largest cash crop. Further field testing confirmed the achievement, and the company began planning full commercialization. But bringing a new pesticide to market was no longer a quick or easy process. New laws required that pesticides be registered and listed with the EPA before marketing. The EPA required submission of extensive data to demonstrate the chemical's efficacy, its safety relative to a broad range of toxicological tests, and the absence of potential environmental effects. Rohm and Haas submitted twenty-four volumes of data on RH-6201 in December 1978, knowing it would be forced to wait one to two years before it could expect EPA action.

In the meantime, the firm had begun marketing the product, now named Blazer, in countries such as Argentina and Brazil, which had large soybean crops and less cumbersome regulatory procedures. Blazer had great success in these countries, as well as in six American states where it was used in 1979 under an emergency permit procedure. By April 1980, when the EPA gave its approval to Blazer, the Agricultural Chemicals Business Team was prepared to introduce the herbicide immediately and take advantage of the growing season then getting under way. It unleashed both a large technical field sales effort and an advertising campaign that took the largest share of Rohm and Haas's 1980 advertising budget. The introduction was a great success. In its first year of full commercial exploitation, Blazer became Agricultural Chemicals' largest domestic product line, surpassing both Stam and Dithane. Sales continued to increase over the next several years as Blazer further penetrated the soybean crop market.[3]

Blazer was so successful the company decided to build a new plant for production of the herbicide. Economic factors favored construction in Texas. Rather than add an agricultural chemical line at the Houston plant, Rohm and Haas acquired a new site in nearby La Porte in 1980. Construction began on a facility which would have two production lines: a unit for Blazer and one for the company's growing line of Rocryl specialty acrylic monomers. This new Bayport plant began producing Rocryl in February 1982, and Blazer ten months later.

Kathon biocide was another product to come out of research in the late 1970s. It was first marketed in Europe in 1976 for preventing growth of microorganisms in metalworking fluids and cooling tower water. American introduction followed the next year, after the product gained EPA registration. Rohm and Haas soon began testing Kathon as a preservative for toiletries and cosmetics and, in 1981, introduced a formulation, Kathon CG, for that purpose. Kathon CG has gained wide use in several applications, most notably in shampoos. Biocides were an old market for Rohm and Haas, going back to the 1940s and the Hyamine line pioneered by Herman Bruson. But Kathon, based on completely different chemistry, was a major breakthrough. It replaced formaldehyde-based biocides which had come under attack as environmental hazards. It also proved useful in a far broader range of applications than virtually any competitive product.

The Polymers, Resins and Monomers (PRM) Business Team continued to increase the markets for Rhoplex aqueous acrylic emulsions. Environmental concerns forced many industries to seek alternatives to solvent-based coating systems, and Rohm and Haas research responded with water-based emulsions for industrial coil coatings, roof mastics, and marine applications. Continued improvements in Rhoplex emulsions for trade sales (house paints) brought obtainable gloss levels ever closer to

Rohm and Haas opened its newest plant in La Porte, Texas, in 1982. It became the sole manufacturing location for Blazer herbicide.

those of oil paints, and finally enabled the company to challenge the older technology's dominance in high-gloss applications. Another sustained research program in trade sales led to the 1982 introduction of Rhoplex AC-829. It was the first product that combined the needed degree of chalk adhesion with the other virtues (film build, flow, and hardness) which had made AC-388 the company's top selling paint emulsion for fifteen years.

In a related area, PRM introduced Ropaque OP-42 in 1982. OP-42 was a replacement for a portion of the titanium dioxide which served as the opacity-imparting ingredient in paint. It consisted of tiny uniform polymer spheres whose hollow centers gave them the ability to scatter light. It was the product of an eight-year exploratory research program in the traditional Rohm and Haas mold: a combination of a well-understood technology (acrylic polymers) and a customer group (paint manufacturers) with a novel insight (trapped air voids in a film scatter light) that led to a product for a new application. Although it had been a speculative project, with no guarantee of technical success, it could be justified under RONA guidelines because the potential market and margin for a success would be substantial.[4]

Acryloid processing aids and impact modifiers for use in the manufacture of rigid polyvinyl chloride (PVC) plastics were another product line which performed extremely well in this period. Like many Rohm and Haas products, most of these Acryloids were acrylic polymers (although some were more complicated systems containing methacrylate, butadiene, and styrene). The company had been making these modifiers since the late fifties, but it was only in the late seventies and early eighties that the line became a major one, with sales growth that rivaled Blazer. Acryloid sales grew because the use of PVC grew. It was impossible to make finished products of unmodified PVC.

These Acryloid modifiers were another example of the continued vitality of the classic Rohm and Haas strategy: selling advanced technology to other manufacturers for use in their own products—in this case PVC objects such as plastic pipe, house siding, and window frames. The company maintained highly skilled technical sales and application staffs who were experts in PVC product formulations, and who worked with customers to improve the quality of the finished articles. Acryloids accounted for only a small part of the bulk and cost of the finished product, and they lost their own identity in the course of the customer's processing. Rohm and Haas has never manufactured PVC powder, as that is a commodity and not a high-technology specialty.[5]

Not all product lines had growth potential like that of Acryloid modifiers. In some areas the company's strategy was to manage markets for long-term profitability rather than pursue improbable visions of expansion. Plexiglas sheet, Rohm and Haas's best-known product, fell into this category. Not only was the market for acrylic sheet a mature one, but acrylic sheet was acquiring characteristics of a commodity chemical business. Both domestic and foreign competitors had entered the acrylic sheet market. Often, the competition was able to undersell Plexiglas sheet. Competitors did not support large technical sales service forces. In some cases, they were producing lower-quality sheet through less expensive and less labor-intensive processes than the Rohm and Haas cell casting lines. Many acrylic sheet users found this cheaper sheet was good enough for their purposes. Imports from Taiwan bedeviled Rohm and Haas, as they did many American manufacturers. Rohm and Haas fought back by introducing its own less expensive sheet, Plexiglas MC, made by a new continuous-extrusion process. It also reduced the size of its innovative technical support operation, as this level of overhead was not cost-effective in a commodity market.

Other major products, including Plexiglas molding powder, while occupying market niches of similar maturity to sheet, seemed better positioned for current and future profitability. Molding powder sales remained tied to domestic automobile production, and thus fluctuated with the cyclical vagaries of Detroit's car making.

Under the RONA-based strategy adopted in 1978, Gregory and his staff reassessed each of Rohm and Haas's product lines and operations with a view toward selling or discontinuing those which seemed poorly positioned for future profits. Similarly, they weighed possible acquisitions in cases where purchases could have a synergistic effect on existing business. The company divested product lines deemed uncompetitive or a poor fit with the company's major activities—now explicitly defined as the exploitation of polymer design and agricultural chemistry.

Among the products sold were Paraplex and Monoplex plasticizers, a line of specialty enzymes, Paraplex molding resins, and Hyamine biocides. Some entire businesses were sold, most notably Consolidated Biomedical Laboratories (CBL) in 1982, a group of medical diagnostic laboratories which Rohm and Haas had put together since 1969. CBL was profitable, but it was too far removed from Rohm and Haas's main interests in specialty chemicals, and not a good candidate for further investment.[6]

The first of the new wave of acquisitions actually preceded the formal adoption of RONA. In 1975 Rohm and Haas purchased a Kensington, Connecticut, polycarbonate plastic sheet plant and associated technology from Rowland Inc. Transparent sheet of polycarbonate had begun to challenge acrylics in some of the latter's established applications, especially where superior breakage resistance was important. Among the growing uses of polycarbonate was glazing for schools, buses, and other facilities where the risk of damage was high. Rohm and Haas renamed the Kensington plant's product Tuffak and assigned it to the Plexiglas sheet marketing staff. Tuffak was a logical extension of a market in which Rohm and Haas was established.

In 1981 Rohm and Haas made a larger acquisition along the same strategic lines when it acquired a PVC modifier plant in Grangemouth, Scotland, from the Borg-Warner Corporation. The Grangemouth plant manufactured a line of methacrylate-butadiene-styrene-type PVC modifiers via a different route than that used in Rohm and Haas's plants. The Grangemouth modifiers had unique properties which made them valuable in the manufacture of certain types of PVC plastics, such as transparent bottles, where Rohm and Haas's indigenous modifiers would not work. Along with the plant and its technology, Rohm and Haas obtained a cooperative research agreement with Kureha Chemical Industry Company, the Japanese firm which had developed the technology used at Grangemouth. Rohm and Haas opened an additional facility for Kureha-type modifiers at its Louisville plant in 1984 to serve the domestic market.[7]

Several other purchases fit the pattern set by the polycarbonate and modifier acquisitions. In 1984 Rohm and Haas acquired Hydranautics, developer of a reverse osmosis process for water purification. Management saw this as a natural extension of its longstanding ion exchange technology for water purification. In the same year, the company purchased Duolite International, a French-based subsidiary of the Diamond Shamrock Corporation that was a manufacturer of ion exchange resins. It also acquired a Diamond Shamrock ion exchange plant in California.

The RONA philosophy, and the increased attention given to foreign operations by establishment of the regional matrices, led to a critical reexamination of many overseas operations.

This, in turn, led to several reorganizations and retrenchments. Probably the most painful decision was the one made in 1981 to close the Teesside, England, monomer works, a modern plant that had been opened with great publicity less than a decade before. The cost of operating the relatively small facility was higher than the cost of shutting it down, and the company calculated that Europe could be supplied with acrylate and methacrylate monomers more cost-effectively from the mammoth facilities in Houston.[8]

In several regions and subregions, the reevaluation led to a more widespread retrenchment. Entire subsidiaries and their associated plants were closed in Argentina, Venezuela, and Indonesia. An acrylic sheet plant in Colombia was sold, while investment was increased in other more profitable parts of the Colombian subsidiary, such as pesticide manufacture and sale. An ion exchange resin plant opened in Taiwan in 1979, only to close four years later. A reorganization plan for the European Region announced in early 1983 cut regional employment by one-fourth. Not every change in overseas operations was a retrenchment. In 1981 Rohm and Haas announced that a Kathon biocide plant would be built at the firm's Jarrow, England, works. When completed in late 1983, the plant became the company's sole facility for production of this growing product. For the first time, Rohm and Haas was filling domestic demand for a major product with a facility it had built abroad. Rohm and Haas had taken one more step toward becoming truly multinational. A similar decision to centralize worldwide Dithane production in Lauterbourg, France, followed.

In the 1980s, as plentiful supplies of oil and generally stable raw material prices helped put an end to the worldwide recession, Rohm and Haas enjoyed robust health. It had shifted from a high-debt position to a reserve of cash for investment. In 1983 the company achieved its announced goal of ten-percent RONA, a substantial advance over the seven-percent RONA of five years before. The firm also began a series of quarters of record earnings in 1983 which extended into the following year.

Much of the investment capital went into continued research and support of the more attractive areas of its core businesses. From the Research Division came product ideas to reach previously untapped markets that demonstrated the continued vitality of acrylic chemistry. Among these were a range of adhesives and a polymer concrete system designed to aid in the repair and restoration of roads and bridges.

As the Rohm and Haas Company began its seventy-fifth

year, it had ample reason to believe it was well poised for the future. The matrix organization provided the necessary channels for coordinating corporate control with a degree of decentralized decision making. The RONA principle provided a guideline for managerial decisions. It became the means by which Gregory and his executive team implemented the classic Rohm and Haas strategies in the modern era. Management used RONA to help it decide where it should invest in new product research, where it should seek acquisitions, and where it was necessary to retrench.

RONA was an invaluable tool in Rohm and Haas's recovery from the lean years of the mid-1970s, and it was a potent procedure for managing ongoing operations. But by itself, it was unlikely to provide the blueprints for future directions. It was, by its nature, a relatively short-term technique. As Gregory prepared to lead the company toward the celebration of its diamond jubilee, thoughts of what would come in the years ahead began to occupy a greater percentage of his, and his management team's efforts.

NOTES

1. "Rohm and Haas Company Financial Analysts Meeting, October 14–15, 1975." "Rohm and Haas Company, Annual Stockholders' Meeting, April 24, 1978, Comments by Vincent L. Gregory, Jr., President," copies in Rohm and Haas Company archives.

2. Richard Rosera, "Developing a Rodenticide: The Life and Death of Vacor," unpublished paper, copy in Rohm and Haas Company archives.

3. "The Birth of a Herbicide," Agrichemical Age, March 1980, pp. 6–8, 58. Rohm and Haas Company, "Annual Report for 1980," p. 3, "Annual Report for 1981," pp. 2, 11.

4. Richard Harren, conversation with the author, August 10, 1984. Richard Harren, "Elements of a Successful Research Project: The Development of an Opaque Polymer," Journal of Coatings Technology, 55 (December 1983): 79–81.

5. Charles Pyle, conversation with the author, October 16, 1984.

6. Gregory interview.

7. Pyle conversation. Barbara Rudolph, "It's the Little Things that Count," Forbes, September 12, 1983, p. 175.

8. Gregory interview. Rohm and Haas Company, "Annual Report for 1981," pp. 2, 6.

14

ROHM AND HAAS
AT SEVENTY-FIVE

Rohm and Haas celebrated its seventy-fifth anniversary in 1984 with a banner year. Company sales passed $2 billion for the first time, and earnings reached a new record level of $172.2 million, a twenty-five percent increase over 1983, which itself had been a record year. RONA increased again, reaching twelve percent, far surpassing the goal set in 1978. The firm's performance received favorable notice in the business press. *Forbes* magazine's "37th Annual Report on American Industry" ranked Rohm and Haas at the top of the chemical industry in both Return on Equity and Earnings per Share Growth, the publication's key business performance measures.[1] Much of the chemical industry enviously eyed Rohm and Haas's emphasis on high-margin specialty chemicals. Other firms began to follow its lead, switching to specialty operations from commodity chemicals, which seemed plagued by slow growth, low profits, and cyclicality.[2]

The pace-setting company that enjoyed this praise and high standing was a varied, yet focused one. It operated plants in nineteen countries around the world and, with thirty-one percent of its sales outside North America, was a flourishing multinational enterprise. It engaged in many businesses, but all of them were linked by a mission that Gregory described in the fall of 1984. He said Rohm and Haas's mission was to "exploit our expertise in polymer design and to develop agricultural and specialty industrial chemicals." The company would continue to seek success in much the same ways that it had since the days of its founder.

Rohm and Haas maintained staffs of technical experts

Vincent Gregory, chairman of the board

around the globe. From an agricultural field representative demonstrating Dithane to coffee growers in Colombia, to a coatings chemist showing the latest high-gloss Rhoplex emulsions to paint manufacturers in Australia, to a leather expert working with tanners in the old leather center of Gloversville, New York, these men and women demonstrated Rohm and Haas's enduring belief that service to customers was a key component of success.

The company continued its unwavering commitment to research as the source of advances in technology. Research activities were centered in the dozen buildings of the Spring House, Pennsylvania, research campus, but additional activities took place at the Bridesburg and Bristol plants, a farm in Newtown, Pennsylvania, and at a European Region Research Laboratory in Valbonne, France. Altogether, the company spent $109 million on research in 1984. That was 5.4 percent of sales, far above the industry average, and higher than the 4 percent which had prevailed at the company in previous years. The Research Division employed 1,500 people.

The continuing involvement of the Haas family remained a steadying influence throughout this period. Although retired from active management, both of Otto Haas's sons continued to sit on the board of directors. John Haas held the titular post of vice chairman of the board, and maintained an office in the Independence Mall headquarters. He could be found there virtually every day, serving informally as a conscience of the company, actively practicing his long-held concern for the well-being of the people of both Rohm and Haas and the Delaware Valley.

Through a combination of direct ownership, stock owned by the William Penn Foundation (a philanthropic institution established by Otto Haas and whose board chairman was John

Vice Chairman John Haas has long been interested in the well-being of the citizens of the Delaware Valley. He is here surrounded by members of the Boys and Girls Club, a charity to which Rohm and Haas and the William Penn Foundation have both long contributed.

Haas), and stock held by several charitable and beneficial trusts, the Haas group still controlled 47 percent of the company. Such a concentration of ownership was extremely unusual for a corporation of Rohm and Haas's age and size. Between the foundation and the trusts, over one-third of the company's dividends flowed to philanthropic channels, primarily in the Philadelphia area. In October 1984 the company agreed to purchase from the foundation 1.33 million shares of its stock (approximately 5.5 percent of the shares outstanding), and announced that it would purchase up to 1.5 million additional shares on the open market. The company–foundation agreement gave both parties the right to repeat the process. Completion of these maneuvers would decrease the quantity of Rohm and Haas stock outstanding by 21 percent, increase the earnings-per-share of the remaining stock, and maintain the Haas group holdings in the company at 47 percent.

The constancy of factors such as Haas family involvement, research effort, and concern for customers, did not mean that the Rohm and Haas organization was static. The company continued to refine its matrix management system and extend its principles further down into the hierarchy. The four business teams developed in the mid-seventies were gradually reorganized into sixteen smaller units. This change gave decision-making responsibility to managers who were closer to customers and markets. The board of directors evolved from a management-dominated body to one containing a majority of outside directors. These outside directors, chosen for their broad range of experience in various aspects of American business and society, served as a pool of independent analysts and consultants on management plans.

The cash flow generated in the 1980s exceeded the amounts Rohm and Haas needed to repurchase stock, finance increased research, and fund new programs. For the first time since the early 1960s, the company was able to consider diversification. Before making any moves, however, it had developed new systems for strategic planning.

The idea of establishing a separate, integrated planning function had a slow beginning at Rohm and Haas. Company planning during Otto Haas's tenure was done by the founder himself. Through the 1960s, planning was a relatively informal part of overall responsibility in F. Otto Haas's team management approach. Late in the decade, the company made its first attempt at writing a five-year plan, but the document was simply a financial forecast generated by the individual business units. It consisted largely of projections of current trends. This type of bottoms-up projecting persisted through much of the 1970s. The plans assumed that high growth would continue. No thought was given to the possibility that external events (such as the OPEC-created oil crisis) might affect the company's future. Gregory established a planning committee in the early 1970s, but it proved to be largely a discussion group rather than one that exerted substantial influence on the company's direction.[3]

The first finding of the company's 1975 Organization Study Group Report was that the existing business planning process was "defective" and inflexible. It cited a "lack of clearly stated business strategies based upon comprehensive market research," and it recommended a matrix structure that incorporated improved provisions for planning. The new Corporate Business Department, as chief integrator of the company's activities, would have primary responsibility for coordination of the company's business plans. Its staff would include a full-time planning manager. A new Management Committee, consisting of Gregory and a handful of key executives, would have responsibility for reviewing and approving long-range plans. With the adoption of this matrix system, planning was established as an important, integral corporate function.[4]

The collapse of the Fibers Division further delayed the development of planning. It created an atmosphere of crisis management in which there was little or no time for looking beyond immediate problems. As the company returned to profitability, however, it gradually shifted its planning emphasis from simple numerical projections to strategic thinking. Increasingly, more attention was devoted to discussions of where Rohm and Haas might be in three or ten years, and how it might get there. A Corporate New Ventures group was formed in 1978 to seek out new opportunities in both current and additional areas, and as money allowed, to investigate acquisitions of both technologies and businesses.

As part of this new strategic planning setup, Director of Corporate New Ventures, William Kulik, undertook a study of the entire specialty chemical industry. He divided the industry into fifty-seven individual businesses, and then extracted a short list of four or five businesses which might be viable Rohm and Haas investment targets. These target areas, which included adhesives, electronic chemicals, biocides, and engineering plastics, had several things in common. They were all areas where Rohm and Haas lacked a substantial presence, but where synergism could be expected with the company's strengths in polymer synthesis and agricultural chemistry. In addition, they had attractive projected growth rates and were of total market size that would enable the company to gain a respectable market share with an investment in the neighborhood of $100 million.

After examining this list of possible acquisition targets, management settled on advanced adhesives as the area in which it would place primary emphasis. In 1983 Rohm and Haas purchased two small businesses, the Isochem Resins Company and Furane Products. These acquisitions were described by Gregory as steps taken "to strengthen our position in adhesives."[5] With a third acquisition, the Electro-Materials Corporation of America (a producer of coatings and conductive pastes for electronic components), these three units became the Rohm and Haas Advanced Materials Company (RHAMCO) in the PRM Business Team. However, the individual firms retained their independent identities.

While Rohm and Haas was taking these steps toward building a stand-alone adhesives business, outside circumstances led both Kulik and the Management Committee to look more seriously at investing in electronic chemicals. Electronic chemicals are expensive, high value-added polymers used in the manufacture of electronic components such as microchips and integrated circuits. In late 1979, the Shipley Company, a privately held leading manufacturer of photoresists (chemicals used in the etching of chips and circuits), reached an agreement to be acquired by Du Pont. In May 1980 this agreement fell apart. Rohm and Haas approached Shipley and initiated discussions about a possible friendly takeover. Rohm and Haas learned that the Shipley family, although desiring the influx of capital and research support that alliance with a larger firm would make possible, no longer wanted to sell its company outright. After extended negotiations, Rohm and Haas purchased a thirty-percent interest in Shipley in 1982. It acquired this minority interest in the clear hope that the Shipley family would eventually sell the rest of the company to Rohm and Haas. Kulik moved from his position at Rohm and Haas in June 1983 to become executive vice president of Shipley, further strengthening the growing ties between the two firms. In the two years following the agreement, Shipley continued the rapid profitable growth that first attracted Rohm and Haas interest. Shipley sales reached $120 million in 1984.[6]

In 1984 Rohm and Haas acquired an additional chemical

supplier to the electronics industry, Plaskon Electronic Materials Incorporated, a manufacturer of materials used in the packaging and assembly of electronic components. This completed its shift in acquisition emphasis from adhesives to electronic chemicals. As the primary market for the adhesives made by Isochem and Furane was the electronics industry, the five acquisitions were grouped together in 1984 and spoken of as an electronic chemicals business. Including all of Shipley, the five companies had sales of $205 million in 1984. RHAMCO was moved from PRM to Industrial Chemicals. As 1984 ended the companies continued to be operated as distinct entities. Rohm and Haas management saw the links between the individual firms (and between the group and the parent company) in research and technology, rather than in manufacture and marketing. In a span of just two-and-one-half years, Rohm and Haas had become a major player in the several markets that together made up the electronic chemicals industry.

It was to encourage the managers of these acquired businesses to be more entrepreneurial, to be willing to take higher risks, and to steer clear of much of the bureaucracy of the parent corporation that Rohm and Haas retained their individual corporate structures. Electronic chemicals, although the most prominent and largest investment, was not the only area where Rohm and Haas used decentralization as a technique for exploiting a new, high-potential market. Rohm and Haas Seeds Inc. was formed in 1983 to commercialize a long-developing company research effort in gametocides, chemicals which destroy male fertility in plants without affecting female fertility. This permits agronomists to use hybridizing techniques to develop improved strains of wheat. Rohm and Haas acquired several small seed wheat and soybean businesses, and began marketing improved hybrid seed under the tradename Hybrex.

As the calendar turned from 1984 to 1985, the Rohm and Haas Company completed its seventy-fifth year of existence. Business was thriving. The company celebrated its milestone with an air of well-deserved self-congratulation. It was a company with a history but it was also a company preparing for its future, poised with plans which management believed would provide foundations for achievements to come. The Rohm and Haas "Annual Report for 1983" (published in March 1984) had featured an essay on the company's history. The companion volume for 1984 contained a special section on the firm's new technologies. Rohm and Haas had changed immeasurably over its seventy-five years. But a modern observer might conclude that, if Otto Haas returned today, he would find a company of which he could be proud, a company which continued to fulfill the dream which he had dreamt so many years before.

NOTES

1. "Chemicals," Forbes, January 14, 1985, p. 118.
2. "The Urgent Rush to Specialty Markets," Chemical Week, October 28, 1983, pp. 40–52.
3. J. L. Wilson, conversation with the author, February 21, 1985.
4. C. J. Prizer, "Organization Group Study Report," May 1, 1975, copy in Rohm and Haas Company archives.
5. Rohm and Haas Company, "Annual Report for 1983," p. 3.
6. Wilson conversation.

CONCLUSION

Over the course of seventy-five years, Rohm and Haas Company has grown from a small, single-product storefront operation to a major corporation with plants and offices around the world. The passage of time and the process of growth have made it a very different creature from what it was in the beginning. Yet, through all of the changes, through all of the development, there has been constancy at the core; a certain set of central principles has given the firm a consistent direction.

Companies, like people, can have an enduring set of inherent values from which they develop characters of their own. In the case of Rohm and Haas, these values centered on the business style Otto Haas developed while he was on the road selling Oropon. Haas approached the tanners as a highly skilled technician with an innovative, scientifically advanced product, one which would lose its own identity in the course of manufacture. He made his mark by helping to solve problems in the tanneries. In the following decades he sought to sell other products the same way. Time and again, he succeeded with syntans, textile chemicals, coatings ingredients, and oil additives.

His commitment to such scientifically advanced products had several important corollaries. Instead of manufacturing consumer goods or commodities, he focused on technical innovation as a key business strategy. He selected his marketing personnel for their technical training and ability, rather than sales experience. He kept a low public profile both for himself and his company. He applied his basic strategy even to Plexiglas sheet, the most important product in Rohm and Haas's history. In 1945 someone less committed to Haas's principles might have been tempted to leave in place the large Plexiglas fabrication facilities which the company had operated as part of the war effort. But Haas wasn't interested in making finished parts, nor did he wish to compete with potential customers. As a result, he abandoned Plexiglas fabrication as quickly as possible and instead maintained a staff of technical experts to help develop new uses for Plexiglas. The strategy succeeded. Working with sign makers, architects, and lighting fixture manufacturers, Rohm and Haas established Plexiglas as a new material with unique properties. It became the product of choice for the manufacture of a large range of goods.

As the company grew, Haas found it impossible to control all operations personally, as he had in the past. What he did, perhaps without realizing it, was the next best thing. He picked managers from within the company who had been schooled in his business principles and adhered to them. Having grown up in the system, they followed the founder's guidelines.

Otto Haas retired in December 1959 at the age of eighty-seven, and died the next month. His son, F. Otto Haas, then headed the company in a critical transitional period. By this time, Rohm and Haas had moved well past the entrepreneurial stage and F. Otto Haas's managerial style took this into account. During his decade in charge, he devoted much time and attention to developing a formal organizational structure more appropriate for the size and complexity that characterized Rohm and Haas in the 1960s. The development of this new structure had to occur if the company were to survive.

At the same time, he sought to match his father's successes in developing new business. Most notable was the effort to develop a novel elastomeric fiber spun from acrylic emulsions for the textile industry. But his management team let their enthusiasm for fibers divert them from their company's time-tested strategies. Managers who had spent decades successfully selling specialty products and technical expertise proved incapable of marketing products which differed little from those of their competitors except in price. Imprisoned by their training, they kept trying to invent specialties where a market for them did not exist in an industry geared to using only a few versatile fibers.

The company's older specialty chemicals businesses still pursued the founder's type of market niches, however. And these businesses were profitable enough to offset the losses from fibers. In at least one case, Rhoplex paint emulsions, the company conquered a new market.

F. Otto Haas retired in 1970. His successor, Vincent Gregory, completed Rohm and Haas's difficult transformation from a paternalistic family-owned business to a large multinational corporation run by professional managers who lacked a major stake in ownership. Professional tools such as strategic planning, RONA, and an independent board of directors became part of the business practice. Initially, Gregory expanded the firm's fibers operations but then discontinued them with heavy losses in 1976. Gregory and his management team reembraced the founder's strategies which were now well established in the Rohm and Haas corporate culture and found in them a base for greater sales and profitability as the company developed biocides, herbicides, roof mastics, adhesives, electronic chemicals, and other new products. Although the volume of business greatly increased, the values and philosophies of the company under Gregory were those of the founder—direct dealings with the customer, a strong emphasis on research, and concern for employees.

The company's evolution was hardly the result of organizational size and financial resources alone. New science and new technology were primary factors in shaping its development. Otto Haas's own position astride the two cultures of his native and adopted lands provided an important ingredient in his company's success. It placed Rohm and Haas in a favorable position, which Haas ably exploited, as a conduit for the transfer of German chemical technology to the United States. As this conduit closed, internal channels to innovative science and technology developed. The dramatic and sudden growth of Plexiglas; the massive company capacity for manufacture of acrylic monomers and Plexiglas sheet; the Research Division's development of a new, cheaper process for acrylate synthesis (the F Process); the success of acrylic paint emulsions in the late 1960s, all helped frame the key problems and growth opportunities in the history of Rohm and Haas.

Interaction with the United States government, especially the intervention of the Alien Property Custodian during and after the two world wars, fundamentally changed the company's direction. The APC imposed an external agenda on the prerogatives of Otto Haas and the Rohm and Haas management. If not for this intervention, Rohm and Haas might still be a privately held corporation dually based in the United States and Germany.

The United States in the twentieth century has been variously looked upon as existing in the age of the corporation and in the age of science. In reality, as the history of Rohm and Haas shows, it exists in both. Perhaps, the century is best characterized as the age of corporate science. The advancement of business has become dependent on the continual introduction of new science and novel technology. The progress of science has become heavily dependent on corporate employment, corporate facilities, and corporate finances. In the corporate setting, science, technology, and economics are interdependent.

The search for such broadly based patterns does not make the particular any less significant. The first seventy-five years of the Rohm and Haas Company required far more than the playing out of impersonal forces. The work of Otto Haas, his successors, and associates demonstrated that their corporation was equally the product of the decisions and methods of human beings. It was their individual decisions and methods that became Rohm and Haas's corporate values and principles. Through their personal efforts, the company acquired an institutional identity and a continuity that could thrive even a quarter-century beyond the death of a giant such as Otto Haas.

Rohm and Haas Home Office, Independence Mall West, Philadelphia, 1984.

CHEMICAL EQUATIONS AND FORMULAS FOR SELECTED PRODUCTS

CHAPTER 2

Lykopon: Sodium hydrosulfite (sodium dithionite) $Na_2S_2O_4$

Formopon: Sodium formaldehyde sulfoxylate dihydrate
$NaHSO_2 \cdot CH_2O \cdot 2H_2O$

Formopon Extra: Basic zinc formaldehyde sulfoxylate
$Zn(OH)HSO_2 \cdot CH_2O$

CHAPTER 3

Lethane 384: β-butoxyethyl-β′-thiocyanoethyl ether
$C_4H_9OC_2H_4–O–C_2H_4SCN$

CHAPTER 5

Acrylates: "A" Process:

$HOCH_2CH_2Cl$ + $NaCN$ → $HOCH_2CH_2CN$

ethylene   sodium   ethylene
chlorohydrin   cyanide   cyanohydrin

$\xrightarrow[\text{H}^+]{\text{ROH/H}_2\text{O}}$ $CH_2{=}CHCOOR$

$\dfrac{\text{aq. alcohol}}{\text{acid}}$ alkyl acrylate

Methyl Methacrylate: "A" Process

CH$_3$CCH$_3$ + NaCN → (CH$_3$)$_2$C—CN $\xrightarrow[\text{H}^+]{\text{CH}_3\text{OH/H}_2\text{O}}$ (CH$_3$)$_2$CCOOCH$_3$ $\xrightarrow{\text{P}_2\text{O}_5}$ CH$_2$=CCOOCH$_3$
   ‖ | | |
   O   OH   OH   CH$_3$

acetone   sodium   acetone   aq. methanol methyl   phosphorus methyl
                   cyanide   cyanohydrin acid   α-hydroxyiso- pentoxide methacrylate
                                                butyrate

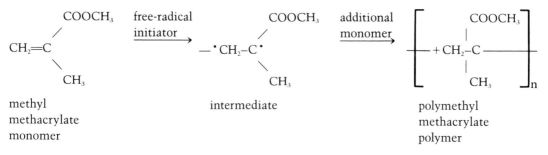

Polymethyl Methacrylate:

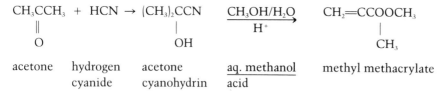

methyl   intermediate   polymethyl
methacrylate   methacrylate
monomer   polymer

CHAPTER 7

Methyl Methacrylate: "B" Process

CH$_3$CCH$_3$ + HCN → (CH$_3$)$_2$CCN $\xrightarrow[\text{H}^+]{\text{CH}_3\text{OH/H}_2\text{O}}$ CH$_2$=CCOOCH$_3$
   ‖ | |
   O   OH   CH$_3$

acetone   hydrogen   acetone   aq. methanol   methyl methacrylate
                   cyanide   cyanohydrin acid

Acrylates: "F" Process

HC≡CH + CO + ROH $\xrightarrow[\text{HCl}]{\text{Ni(CO)}_4}$ CH$_2$=CHCOOR

acetylene   carbon   alcohol   nickel   alkyl
                   monoxide   carbonyl   acrylate
                                            hydrogen
                                            chloride

Acryloid polymethacrylate lubricant additive

$$\left[\begin{array}{c} \quad\quad\quad\text{CH}_3 \\ \quad\quad\quad/ \\ -\text{CH}_2-\text{C}------ \\ \quad\quad\quad\backslash \\ \quad\quad\quad\text{COO[CH}_2]_{11\text{-}17}\ \text{CH}_3 \end{array}\right]_n$$

DDT: p,p′-dichlorodiphenyltrichloroethane (1,1,1-trichloro-2,2-bis(4-chlorophenyl)ethane)

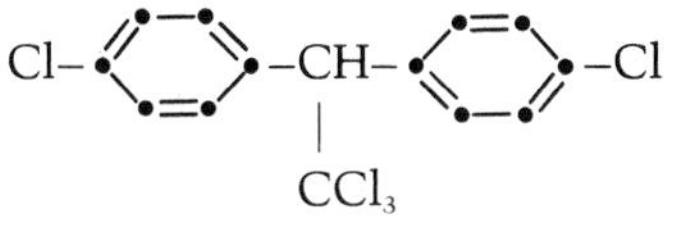

Cl–⟨　⟩–CH–⟨　⟩–Cl
         |
        CCl$_3$

Triton N101 non-ionic surfactant   poly(ethylene glycol)nonylphenyl ether

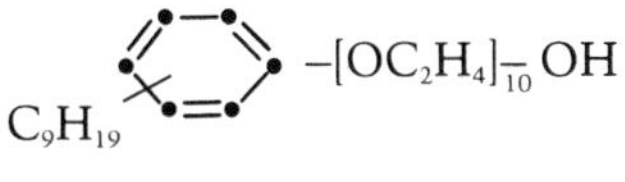

⟨　⟩ –[OC$_2$H$_4$]$_{\overline{10}}$ OH
C$_9$H$_{19}$

Dithane D-14 (nabam): Disodium ethylenebisdithiocarbamate

$$
\begin{array}{c}
\qquad\quad S \\
\qquad\quad \| \\
CH_2NHC\text{–}SNa \\
| \\
CH_2NHC\text{–}SNa \\
\qquad\quad \| \\
\qquad\quad S
\end{array}
$$

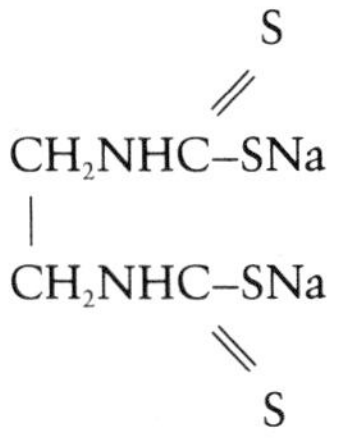

Dithane Z-78 (zineb): Zinc ethylenebisdithiocarbamate

$$
\begin{array}{c}
\qquad S \\
\qquad \| \\
CH_2NHC\text{–}S \\
| \qquad\qquad \backslash \\
| \qquad\qquad\quad Zn \\
| \qquad\qquad / \\
CH_2NHC\text{–}S \\
\qquad \| \\
\qquad S
\end{array}
$$

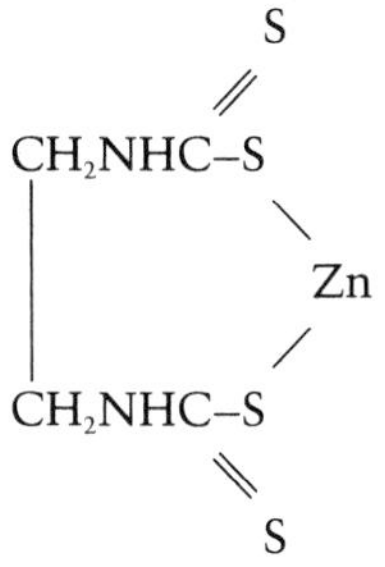

Dithane M-22 (maneb): Manganese ethylenebisdithiocarbamate

$$
\begin{array}{c}
\qquad S \\
\qquad \| \\
CH_2NHC\text{–}S \\
| \qquad\qquad \backslash \\
| \qquad\qquad\quad Mn \\
| \qquad\qquad / \\
CH_2NHC\text{–}S \\
\qquad \| \\
\qquad S
\end{array}
$$

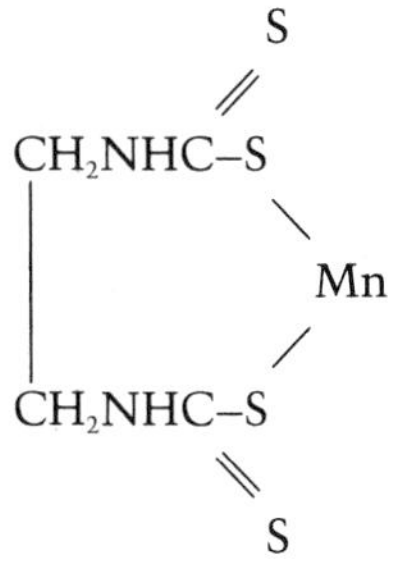

Dithane M-45 (Mancozeb): Zinc ion complex with maneb

Rhothane: 1,1-dichloro-2,2-bis(4-chlorophenyl)ethane

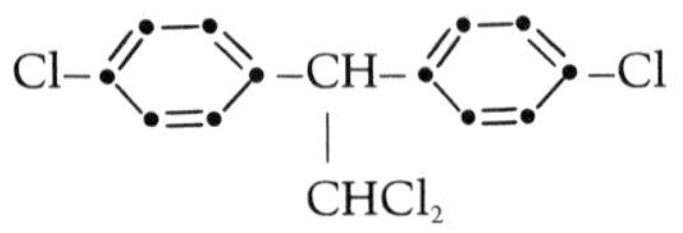

$$Cl-\!\!\!\!\bigcirc\!\!\!\!-CH-\!\!\!\!\bigcirc\!\!\!\!-Cl$$
$$|$$
$$CHCl_2$$

Kelthane: 1,1-bis(4-chlorophenyl)-2,2,2-trichloroethanol

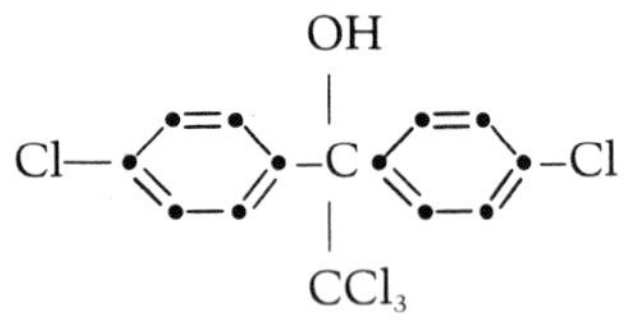

$$OH$$
$$|$$
$$Cl-\!\!\!\!\bigcirc\!\!\!\!-C\!\!\!\!\bigcirc\!\!\!\!-Cl$$
$$|$$
$$CCl_3$$

Perthane: 1,1-dichloro-2,2-bis(4-ethylphenyl)ethane

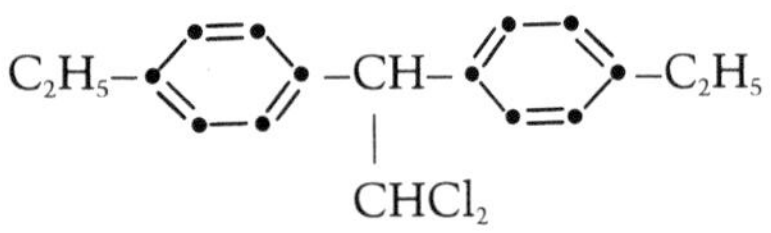

$$C_2H_5-\!\!\!\!\bigcirc\!\!\!\!-CH-\!\!\!\!\bigcirc\!\!\!\!-C_2H_5$$
$$|$$
$$CHCl_2$$

Karathane: mixture of 2,4-dinitro-6-octylphenyl crotonate and 2,6,-dinitro-4-octylphenyl crotonate

$$O_2N \quad\quad C_8H_{17}$$
$$OCCH\!\!=\!\!CHCH_3$$
$$NO_2 \quad \| $$
$$O$$

$$H_{17}C_8 \quad\quad NO_2$$
$$OCCH\!\!=\!\!CHCH_3$$
$$NO_2 \quad \| $$
$$O$$

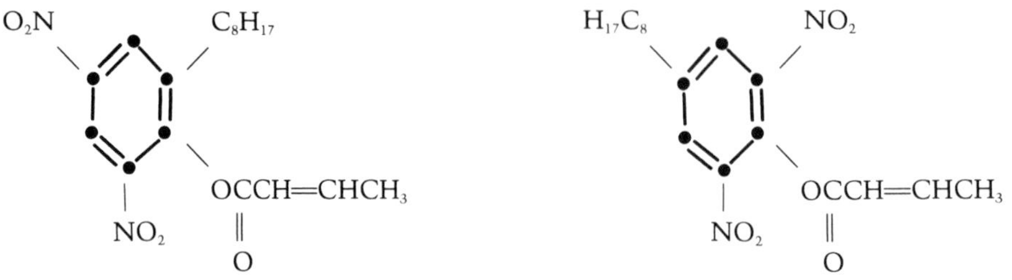

CHAPTER 10

Acrylates: "T" Process

$$CH_2\!\!=\!\!CHCH_3 \ + \ O_2 \xrightarrow[\text{air}]{\overset{\Delta}{\text{catalyst}}} CH_2\!\!=\!\!CHCOOH \ + \ \xrightarrow[\text{H}^+]{ROH/H_2O} \quad CH_2\!\!=\!\!CHCOOR$$

propylene   acrylic acid   <u>aq. alcohol</u>   alkyl
                                                    acid   acrylate

Stam: N-3′,4′-dichlorophenylpropionamide

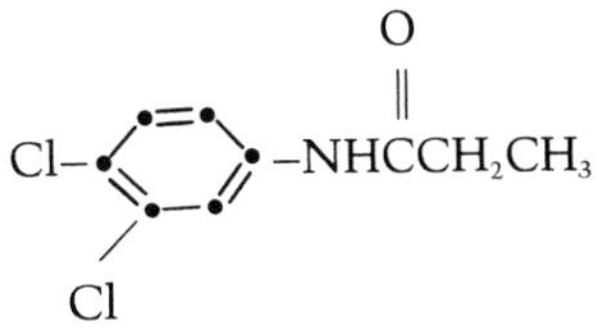

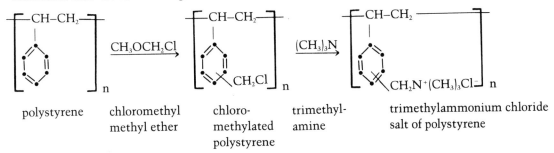

Cl–⬡–NHCCH₂CH₃ (with O double bond), Cl substituent below

TOK: 2,4-dichlorophenyl-4′-nitrophenyl ether

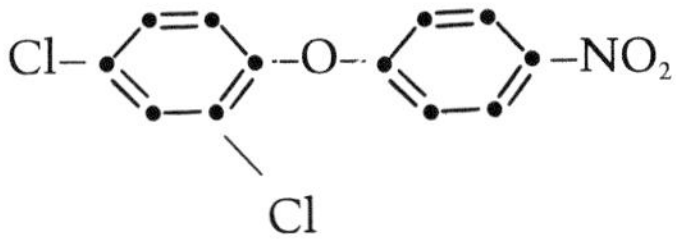

Cl–⬡–O–⬡–NO₂, Cl substituent below

CHAPTER 12

Amberlite IRA-400 ion exchange resin

polystyrene   chloromethyl   chloro-   trimethyl-   trimethylammonium chloride
               methyl ether   methylated   amine   salt of polystyrene
                               polystyrene

ROHM AND HAAS COMPANY TRADEMARKS

Trademarks	Principal Application	Date of First Use (or Registration)
ACRYLOCOAT	Binders and stickers	December 5, 1977
ACRYLOID	1. Synthetic resin forming transparent film	September 14, 1934
	2. Chemicals for lubricants and hydraulic fluids	September 16, 1940
	3. Synthetic resin for inks, coatings, films, and adhesives	September 14, 1934
	4. Solutions for synthetic organic thermoplastic resins and lacquers	September 14, 1934
	5. Synthetic polymeric materials to modify plastic manufacturing	1957
	6. Synthetic resin powder/solution used as additives in manufacturing	September 14, 1934
ACRYSOL	1. Water emulsion of synthetic resins	October 3, 1934
	2. Synthetic resin solution for industrial arts	April 5, 1944
	3. Water emulsions	January 9, 1981
AMBERGARD	Resins and adsorbents for water purification	December 7, 1977
AMBERLITE	1. Synthetic or chemically prepared resins	May 21, 1927
	2. Material to remove ions or acid from fluids	November 2, 1940
AMBERLYST	1. Catalysts	June 20, 1960
	2. Ion exchange resins for industrial arts	January 12, 1962
AMBEROL	1. Synthetic or chemically prepared resins	May 16, 1924
	2. Synthetic resins for the preparation of printing ink and rubber	May 16, 1924
AMBERPLEX	Material to remove ions and acid from fluids	August 28, 1952
AMBERSEP	Inert polymer beads for ion exchange	December 18, 1979
AMBERSORB	Synthetic adsorbents, especially carbonaceous adsorbents	April 12, 1977
AMBERTHERM	Ion exchange and thermally regenerable resins	December 14, 1977
AMBORANE	Polymer reduction resins	March 26, 1979
ANALYTICS	Biomedical and environmental testing services	
ANIM/8	Elastomeric fibers, continuous filaments	May 22, 1969
ANIMATE	Elastomeric fibers, continuous filaments	August 1, 1969

Trademarks	Principal Application	Date of First Use (or Registration)
AQUAPLEX	Resinous Products Co. trademark for water-soluble coating resins	August 27, 1935
ARAZYM	Proteolytic enzyme of pancreatic origin used as a dehairing agent on hides	October 31, 1916
AUTOSERVE	Service program, biomedical sciences	
AUTOPAK	Chemical reagents	March 16, 1971
AYRCRYL	1. Synthetic fibers 2. Latexes 3. Acrylic-coated fabrics	April 27, 1967 October 18, 1977 January 9, 1978
AYRLYN	Nylon and continuous fibers for industrial arts	November 18, 1966
B-1956	Chemical used as emulsifying and spreading agent	January 7, 1942
BAKTHANE	Insecticides	Expired November 21, 1981
BIOTEX	Agricultural chemicals for control of mildew, slime, algae, weeds, microorganisms	June 2, 1971
BLAZER	Agricultural herbicides	June 16, 1977
CARODEL	Polymer resin chips	May 20, 1981
CARODEL CORP	Polymer resin chips	September 27, 1976
CBL	Clinical diagnostic services, biomedical testing	Sold to Hoffman-LaRoche, December 1, 1982
CEMENT PS-30	Synthetic resins and solutions for use as adhesives	March 29, 1971
CHROMESAVER	Tanning agents	August 25, 1980
COMBOSTAT	Chemicals for radioassay	February 20, 1981
COMPETE	Herbicides	December 29, 1978
CRYSTALITE	1. Synthetic resins, their solutions or mixtures 2. Acrylic resin molding powders. Name abandoned in favor of PLEXIGLAS in 1943	August 24, 1926 July 6, 1937
DESAL	Ion exchange resins	September 1970
DIALEC	Toner base polymers	January 16, 1976
DIAMER	Synthetic resins	March 14, 1980
DIKAR	Fungicidal and miticidal preparations	March 13, 1968
DITHANE	Insecticides, fungicides, disinfectants	April 28, 1944
DMP	Substituted aminomethylphenols	November 25, 1946
DR	Acrylic plastic molding materials	June 18, 1974
DUOLASE	Immobilized enzyme preparations for industrial and technical uses	February 1, 1974
DUOLITE	1. Zeolites for filtration and purification 2. Ion exchange resins	April 29, 1943 January 2, 1973
DURAK	Sheets of synthetic plastic film	June 4, 1969
DYTOL	Aliphatic alcohols	May 21, 1953
EPIBOND	Synthetic resins used in adhesives	November 29, 1954
EPOCAST	Casting, bonding, and laminating resins	January 1954
EX/STATIC	Liquid to prevent static electricity	May 10, 1955
EX/STATIC PLASTI SHINE	Cleaning and polishing agent for plastic, enamel, and porcelain surfaces	January 16, 1957

Trademarks	Principal Application	Date of First Use (or Registration)
FLAIR	Colored and uncolored plastic sheets	August 16, 1966
FORE	Fungicides	February 1, 1965
FORMOPON	Sulphoxylate of sodium formaldehyde	August 1, 1919
FURA-PACK	Adhesives for industrial uses	February 1, 1969
GOAL	Herbicides	May 13, 1970
HPF	Coating compound with colorants for leather	March 1, 1973
HYAMINE	Insecticides, fungicides, disinfectants	Sold to Lanza, Inc., 1984
HYBREX	Hybrid wheat seed	October 21, 1982
HYBREX HIGHLIGHTS	Newsletter for hybrid seed agriculture	April 1984
HYDRHOLAC	Lacquer emulsions for leather finishing	April 1, 1947
IMPLEX	Synthetic resin molding powder sold as OROGLAS outside the Western Hemisphere	December 17, 1956
INDAR	Plant fungicides	June 14, 1974
ISOFLEX	Automatic gamma ray counters	May 21, 1982
JET-KOTE	Resin compounds and mixtures	January 1948
KAMAX	Thermoplastic molding resins	November 14, 1978
KARATHANE	Insecticides, acaricides, fungicides	January 30, 1950
KATHON	1. Herbicides 2. Fungicides 3. Synthetic organic chemicals used in biocides	May 31, 1949 December 4, 1963 September 13, 1973
KELTHANE	Insecticides, fungicides, disinfectants	November 16, 1955
KERB	Herbicides	August 18, 1969
KORAD	1. Synthetic plastic film 2. Molding and extrusion compounds, sheets of synthetic resin film	Sold to Korad December 12, 1975 Sold to Korad February 9, 1977
KOREON	Chrome tans	October 31, 1916
KYDENE	Synthetic resin material in bulk form and molding powder	November 7, 1969
KYDEX	Sheets of synthetic resin material	February 28, 1964
LETHANE	Insecticides, fungicides, disinfectants	June 12, 1928
LEUKANOL	Hide tanning chemicals	August 1, 1923
LEUKOTAN	Synthetic tanning preparations	December 18, 1979
LUFAX	1. Zirconium opacifier for porcelain enamel 2. Beater additive for use in manufacture of paper	October 27, 1931 April 18, 1967
LUSTRONE	Waterproof lacquer for use on leather	June 23, 1928
LUTRON	Chemicals to prepare hides for tanning. Same product as OROPON, sold in Africa and Europe	August 19, 1947
LYKOPON	Anhydrous sodium hydrosulfite	June 15, 1919
MACC	Data reduction system for biomedical sciences	February 12, 1984
MENDOK	Herbicides	June 24, 1982
MICROMEDIC	1. Biomedical laboratory instrument for examination of blood 2. Reagents for biomedical analysis 3. Automatic microscopial laboratory apparatus	September 2, 1970 September 11, 1970 November 29, 1983

Trademarks	Principal Application	Date of First Use (or Registration)
MICROSTAT	Dental, medical, and surgical instruments	February 25, 1975
MIRACRYL	Polymers for manufacture of cosmetics and toiletries	August 18, 1977
MONOBED	Synthetic organic material to remove or replace ions	March 28, 1950
MONOPLEX	Synthetic organic chemicals for softeners, plasticizers	Sold to C. P. Hall Co. December 11, 1981
NOVAR	Motor oil sludge dispersants	September 28, 1965
OILSOLATE	Metallic salt employed as a varnish drier	
ORNAL	Synthetic resin molding powders	January 31, 1964
OROGLAS	Synthetic resin material available in sheets, rods, and molding compounds. The trademark OROGLAS is used in place of PLEXIGLAS and IMPLEX on goods sold in countries outside the Western Hemisphere.	August 6, 1947
OROPON	Chemicals to prepare hides for tanning. Same product sold as LUTRON in Europe and Africa	March 1908
OROTAN	Dyes, bleaches, tanning compounds, glue color lakes	December 15, 1942
ORTHO-CLEAR	Cellulose-ester lacquers	January 8, 1927
ORTHOCHROM	1. Finish colors in nature of lacquers 2. Thinners for use with various lacquers	May 15, 1925 December 1928
ORTHODULL	Chemical coating compounds for leather	December 28, 1954
ORTHOLITE	1. Water- and abrasion-resistant finishes 2. Water- and abrasion-resistant finishes for leather	March 27, 1947 August 1947
ORTHOZYM	Proteolytic enzyme preparations	March 16, 1934
PARALOID	1. Acrylic ester for inks, coatings, etc. 2. Synthetic resin and liquid solution for plastic compounds	February 26, 1952 February 26, 1952
PARAPLEX	1. Condensation product for lacquers and coatings 2. Synthetic resin for protective coatings	July 7, 1930 July 7, 1930
PERTHANE	Insecticides, fungicides, disinfectants	March 17, 1953
PLASPREG	Synthetic resinous products	January 16, 1946
PLEX	Synthetic resin material in sheets	September 1949
PLEX-I-GLOSS	Plastic floor coverings	September 11, 1959
PLEXENE	Synthetic resin material in form of molding compounds	November 13, 1942
PLEXI	1. Sheets of resinous materials used as glass substitute 2. Synthetic resinous materials in sheets, rods, tubes 3. Sheets of synthetic resinous material used in furniture manufacture	March 1955 March 1955 April 9, 1970
PLEXI-CRAFT	Plastic rods, sheets, and tubes for decoration	April 5, 1974
PLEXI-VIEW	Plastic-based mirrors	October 1965
PLEXICRETE	Polymer concrete mixes	November 1, 1978
PLEXIGLAS	1. Sheets of transparent, resinous material used as glass 2. Plastic sheets for construction	June 5, 1935 June 5, 1935

Trademarks	Principal Application	Date of First Use (or Registration)
	3. Synthetic resinous material in sheets, rods, and tubes	August 27, 1937
	4. Synthetic resin molding powders	September 30, 1943
	5. Sheets of resinous material used in the manufacture of furniture	March 1955
	6. Mirrors with base material made of plastic	July 16, 1971
	7. Acrylic plastic sheets	November 1978
PLEXIGUM	Substance with adhesive and binding properties. The first acrylic product trademark registered by Rohm and Haas	November 5, 1929
PLEXOL	Synthetic organic chemical for petroleum products	October 9, 1947
POLY-EM	Polymeric emulsions	June 1958
POLYGLAS	Plastic sheets	June 1969
POLYGLASS	Plastic sheets	June 1969
PRIMAC	Amine salts for industry	November 12, 1951
PRIMAFLOC	Coagulants for clarifying water and sewage	December 6, 1960
PRIMAL	1. Waterproof coating compound for leathers	June 26, 1933
	2. Synthetic resins for protective coating and bonding agents	February 26, 1952
	3. Pigment dispersions used for leather finishing	December 31, 1953
PRIMAPEL	Soil-resistant polymer for carpets	January 19, 1979
PRIMENE	Amines for industry	November 12, 1951
PRIMID	1. Waterproofing compounds for leather	October 24, 1968
	2. Synthetic solutions for protective coatings	October 24, 1968
PRIMINOX	Industrial amines	November 6, 1951
PS-18	Synthetic resin used as an adhesive	June 25, 1954
RAWHIDE	Herbicides	February 9, 1978
RHONITE	Synthetic resins used as special finish on textiles	May 6, 1947
RHOPLEX	Resin suspension used as:	
	1. Textile finishes	May 20, 1946
	2. Waxes, polishes, starches	December 4, 1946
	3. Adhesives and binders	May 2, 1949
	4. Aqueous acrylic polymer emulsions	1952
RHOTEX	Resinous material used as textile finish and thickening agent	May 1, 1947
RHOTHANE	Insecticides, fungicides, disinfectants	September 25, 1945
RHOTONE	Binder for dyeing and printing textiles	November 16, 1948
ROCRYL	Specialty acrylic monomers	October 15, 1981
ROHM AND HAAS	Trademark used in a variety of plastic, herbicidal, leather, synthetic resinous, and fluid-treating products	October 31, 1965
ROMAX	Oil field chemicals	July 28, 1977
ROMICON	Ultrafiltration synthetic membranes, valves, and tanks	November 27, 1972
ROMID	Synthetic resin used as molding compounds	March 31, 1981
ROPAQUE	Latex emulsions and suspensions	March 24, 1982
ROPET	Molding resins	April 4, 1979
ROWLEX	Extruded synthetic plastic sheet for fabrication and assembly	March 15, 1971

Trademarks	Principal Application	Date of First Use (or Registration)
SABITHANE	Insecticides, fungicides, disinfectants	January 15, 1960
SIMITAR	Herbicides	June 19, 1979
SISTHANE	Fungicides	May 2, 1978
SKANE	1. Mildewcides 2. Mildewcides	June 2, 1971 June 2, 1971
SLASH	Insecticides	March 14, 1979
SORBANOL	Sulfonated anthracene residue employed as a syntan and acquired from *Badische Anilin und Soda Fabrik*	May 1921
STAM	Synthetic organic chemical for use in the growth and/or eradication of plants	August 17, 1959
STAMPEDE	Herbicides	March 1978
STRATABED	Synthetic organic chemical compound for ion exchange materials	May 10, 1965
TAMOL	1. Chemical compounds used by dyeing, tanning, and textile industries 2. Dispersing agent for leather, textile, and other chemicals 3. Dispersing agent for industrial products	July 8, 1925 March 14, 1968 March 14, 1968
TEXTURA	Yarns	August 1, 1954
TEGO	Cementing medium, adhesives	Sold to Goldschmidt June 27, 1978
TITANENE	Textile mordant	October 3, 1916
TOK	Agricultural herbicides used in weed control	December 15, 1962
TRIOBED	Absorbent resins for various industrial uses	June 20, 1979
TRITON	1. Solution of organic salts and cellulose solvents 2. Synthetic organic chemical compound and preparation with various uses	March 9, 1934 March 16, 1935
TUFFAK	1. Plastic sheet 2. All acrylic marking system	September 1, 1960 November 1967
TUFFAK-TWINWAL	Plastic sheets	October 19, 1976
UFORMITE	Synthetic resinous material for bonding agents, surface coatings, ink adhesives	Sold to Reichhold November 22, 1976
VACOR	Rodenticides	Expired June 15, 1982
WOOD-GLAZE	Synthetic resins for sealing, water-proofing	December 1, 1965
WORLD OF ACRYLICS	Loaning of educational films and materials	April 1974
X-STATIC	Yarns, synthetic fibers, and filaments	Sold December 15, 1977 to Sauquoit
XAD	Polymeric and carbonaceous adsorbents for science and industry	June 17, 1977
ZEROLIT	Ion exchange materials	July 15, 1958
ZIRCOTAN	Mineral tanning materials	December 29, 1942

ROHM AND HAAS STOCK
PERFORMANCE 1949-1984

Year	Dividend Per Share		Trading Range[1]			Number of Shares at Yearend Equivalent to 100 Shares of Original 1948 Stock
	Cash	Stock	High	Low	Close	
1948						100.00
1949	$1.00	4%	71.24	36.00	66.50	104.00
1950	1.45	4%	106.00	61.75	98.00	108.16
1951	1.60	4%	160.25	93.87	142.00	112.49
1952	1.60	4%	155.00	110.00	129.00	116.99
1953	1.60	4%	155.87	115.00	149.87	121.67
1954	1.60	4%	291.20	148.00	266.00	126.53
1955	2.00	4%	426.40	257.00	410.00	131.59
1956	2.00	3%	510.00	373.37	379.00	135.54
1957	2.00	3%	423.50	285.00	314.25	139.61
1958	2.00	2%	515.10	312.00	487.00	142.40
1959	3.00	2%	755.82	481.50	730.00	145.25
1960	3.00	2%	780.00	605.00	619.00	148.15
1961	3.00	2%	670.00	500.00	550.50	151.12
1962[2]	0.75	4 for 1	555.00	445.00		604.46
1962[3]	0.75	4%	129.48	73.00	114.00	628.64
1963	1.25	3%	145.00	104.00	124.75	647.50
1964	1.50	3%	165.83	123.00	159.50	666.92
1965	1.60	3%	181.50	151.50	157.00	686.93
1966	1.60	5%	161.00	88.00	91.50	721.28
1967	1.60	3%	120.00	90.75	95.50	742.92
1968	1.60	4%	124.80	73.75	111.00	772.63
1969	1.60	5%	120.00	83.00	85.00	811.26
1970	1.60	3%	91.67	55.50	87.75	835.60
1971	1.60	2%	118.00	85.25	110.50	852.31
1972[2]	0.40	2 for 1	152.75	110.00		1,704.63
1972[3]	0.61	—	91.00	67.50	89.00	1,704.63
1973	1.00	—	116.00	61.50	74.00	1,704.63
1974	1.16	—	94.25	43.62	46.25	1,704.63
1975	1.28	—	86.50	46.50	58.00	1,704.63

Year	Dividend Per Share		Trading Range[1]			Number of Shares at Yearend Equivalent to 100 Shares of Original 1948 Stock
	Cash	Stock	High	Low	Close	
1976	1.28	—	76.75	44.50	48.00	1,704.63
1977	1.28	—	51.50	28.00	32.25	1,704.63
1978	1.40	—	40.25	28.12	31.62	1,704.63
1979	1.76	—	48.50	31.50	48.12	1,704.63
1980	2.16	—	55.25	32.75	46.37	1,704.63
1981	2.56	—	71.75	46.75	61.00	1,704.63
1982	2.80	—	79.12	45.50	78.00	1,704.63
1983[2]	1.40	2 for 1	125.00	75.50		3,409.26
1983[3]	0.80	—	81.00	58.50	60.75	3,409.26
1984	1.80	—	69.25	48.12	63.87	3,409.26

NOTES

[1] Highs and lows in each period have been adjusted to a before-stock-dividend or before-stock-split basis, except where noted. Closing price is not adjusted.

[2] Cash dividend and stock prices on a pre-split basis.

[3] Cash dividend and stock prices on a post-split basis.

NOTE ON SOURCES

The archives and other surviving records of Rohm and Haas Company form the central base upon which this study is built. Some forty boxes of material had been previously designated as archives. The material contained therein ranges from the important (the first company accounting and manufacturing ledgers) to the interesting (at least ten cartons of photographs) to the trivial (an art book presented to Rohm and Haas by a visiting Russian trade delegation). I added approximately sixty boxes of materials retrieved from a warehouse. There is much serendipity about what records have survived the years. One of the most valuable collections is from the corporate secretary's office. Rohm and Haas has had only three corporate secretaries since its 1917 incorporation; fortunately they all had something of the pack rat in them. Although Otto Haas's own copies of the reports he filed on his interwar European trips have disappeared, duplicates were saved by the corporate secretaries. In addition to these archival records, the official bound minute books of the board of directors of the Rohm and Haas Company are kept by the secretary, and these also proved valuable.

Rohm and Haas's own publications form part of the company archives. Among these are annual reports, numerous obsolete technical bulletins, two editions of the firm's previous commemorative volume, *Chemicals for Industry* (Philadelphia: privately printed, 1945 and 1959), and back issues of the company's two magazines, *The Rohm and Haas Reporter* (1944–   ) for customers and *The Formula* (1945–1970, 1979–   ) for employees.

A second well-catalogued and indexed archive exists at the

Rohm and Haas Research Division library at Spring House, Pennsylvania. The research library has microfilm copies of every report written by company scientists going back to those written in 1924 by Harold Turley, Rohm and Haas's first full-time researcher. A cross-indexed card catalog to the reports facilitates both browsing and locating individual items.

Another resource at my disposal was the people of Rohm and Haas, many of whose careers with the company go back a quarter-century or more. Numerous individuals shared their recollections and material from their personal files, and helped me understand the technologies and strategies which have gone into making the modern firm.

A series of oral histories with prominent figures in the company's history was a rich resource of information. Transcripts of these interviews have been added to the Rohm and Haas archives. I was limited in the number of interviews I had time to do, so I chose some individuals who were obvious candidates and others who represented different spheres of Rohm and Haas's activities. The people I interviewed, and their primary company positions were: Ellington Beavers (vice president for research), John Bergin (chief patent attorney), Ralph Connor (vice president for research, chairman of the board), Donald Felley (president and chief operating officer), Donald Frederick (vice president for sales), Vincent Gregory (chairman of the board and chief executive officer), F. Otto Haas (president and chief executive officer), John Haas (personnel director, executive vice president, chairman of the board), Stanton Kelton, Jr. (research chemist, plant manager), Louis Klein (vice president for Resinous Products Division sales), William Kohler (personal attorney to Otto Haas and other members of the Haas family), Eric Meitzner (research scientist), Robert Reitinger (plant engineer), Edward Riener (plant manager), John Schacht (executive secretary to E.C.B. Kirsopp, sales manager), George Schnabel (controller), and Robert Whitesell (vice president for production, vice president for the Chemicals Division).

Trade publications were an important external source. Among the publications consulted, and the years for which they were used were *Leather Trades Review* (1920–1930), *Soap* (later *Soap and Sanitary Chemicals*) (1929–1948), *American Dyestuffs Reporter* (1922–1931), *Paint and Varnish Production Manager* (later *Paint and Varnish Production*) (1931–1977), *Plastics and Molded Products* (1926–1935), and its successor, *Modern Plastics* (1936–1948).

For the World War I years, the records of the Corporate Management Division, Office of Alien Property Custodian, Records Group 131, File CM 1804, General Archives Division, National Archives, Suitland, Maryland, contain much that is no longer available within the company. A small collection of Otto Haas's personal papers and memorabilia is held by the Balch Institute for Ethnic Studies in Philadelphia.

Several reference works taught me the rudiments of particular technologies. Articles in the three editions of the *Encyclopedia of Chemical Technology* (New York: John Wiley and Sons, 1st ed., 1947–1957, 2nd ed., 1963–1971, 3rd ed., 1978–1984) proved most helpful, as did several monographs: Edward Riddle, *Monomeric Acrylic Esters* (New York: Reinhold Publishing Corporation, 1954); Robert Kunin, *Principles of Ion Exchange* (New York: Reinhold Publishing Corporation, 1960); Charles Martens, *Emulsion and Water-Soluble Paints and Coatings* (New York: Reinhold Publishing Corporation, 1960); and Charles Martens, *Waterbourne Coatings* (New York: Van Nostrand Reinhold Company, 1981).

Secondary historical literature for any work such as this begins in two places. For knowledge of the chemical industry in America to 1939, Williams Haynes, *The American Chemical Industry: A History*, 6 vols. (New York: D. Van Nostrand, 1945–1954), is a work that is encyclopedic in scope. For understanding the twentieth century American corporation (and by implication the twentieth century American economic system), the works of Alfred D. Chandler, Jr., specifically his books *Strategy and Structure* (Cambridge, Mass.: MIT Press, 1962) and *The Visible Hand: The Managerial Revolution in American Business* (Cambridge, Mass.: The Belknap Press of Harvard University Press, 1977), are crucial.

A valuable survey of the European chemical industry in the first part of the twentieth century (which covers American companies as well, but within the European context) is provided by Ludwig Haber in his *The Chemical Industry 1900–1930* (Oxford: Clarendon Press, 1971). Robert Friedel's history of celluloid, *Pioneer Plastic* (Madison: University of Wisconsin Press, 1983), contributes much to an understanding of how new materials find their proper roles. John J. Beer, *The Emergence of the German Dye Industry*, Illinois Studies in the Social Sciences, no. 44 (Urbana: University of Illinois Press, 1959), provides excellent background on the rise of the pioneer world-dominating chemical industry. David Noble, *America by Design: Science, Technology, and the Rise of Corporate Capitalism* (New York: Alfred Knopf, 1977), is a thought-provoking attempt at providing a broad superstructure for the themes suggested by its title. Ernst Trommsdorff, *Otto Röhm: Chemiker und Unternehmer* (Düsseldorf: Econ Verlag, 1976), is of tremendous use on the career of Otto Röhm and the German Röhm and Haas Company during Röhm's lifetime.

Published histories are available for a number of other major companies in the American chemical industry. Among these are Don Whitehead, *The Dow Story* (New York: McGraw-Hill, 1968), Dan Forrestal, *The Story of Monsanto: Faith, Hope, and Five Thousand Dollars* (New York: Simon and Schuster, 1977), and William Dutton, *Du Pont: One Hundred and Forty Years* (New York: Scribner's, 1942). None approach the stature of William Reader's massive and scholarly *Imperial Chemical Industries: A History*, 2 vols. (Oxford: Clarendon Press, 1970–1975). Rohm and Haas's interactions with this giant of the British chemical industry have been minor. A number of additional works which proved valuable in understanding and explicating particular portions of the study have been referred to in the notes.